PARTICLES and the UNIVERSE

PHYSICS IN OUR WORLD

PARTICLES and the UNIVERSE

Kyle Kirkland, Ph.D.

PARTICLES AND THE UNIVERSE

Facts On File, Inc.
An imprint of Infobase Publishing
132 West 31st Street
New York NY 10001

ISBN-10: 0-8160-6116-5
ISBN-13: 978-0-8160-6116-7

Library of Congress Cataloging-in-Publication Data

Kirkland, Kyle.
Particles and the universe / Kyle Kirkland.
p. cm. — (Physics in our world)
Includes bibliographical references and index.
ISBN 0-8160-6116-5
1. Physics. 2. Nuclear physics. 3. Quantum theory. 4. Particles (Nuclear physics) 5. Relativity. 6. Cosmology. I. Title.
QC21.3.K535 2007
539.7—dc22
2006018648

Facts On File books are available at special discounts when purchased in bulk quantities for businesses, associations, institutions, or sales promotions. Please call our Special Sales Department in New York at (212) 967-8800 or (800) 322-8755.

You can find Facts On File on the World Wide Web at http://www.factsonfile.com

Text design by Kerry Casey
Cover design by Dorothy M. Preston
Illustrations by Richard Garratt

Printed in the United States of America

MP FOF 10 9 8 7 6 5 4 3 2 1

This book is printed on acid-free paper.

This book is dedicated to Professor George Gerstein, a remarkable scientist and an even more remarkable person.

Contents

PREFACE

THE NUCLEAR BOMBS that ended World War II in 1945 were a convincing and frightening demonstration of the power of physics. A product of some of the best scientific minds in the world, the nuclear explosions devastated the Japanese cities of Hiroshima and Nagasaki, forcing Japan into an unconditional surrender. But even though the atomic bomb was the most dramatic example, physics and physicists made their presence felt throughout World War II. From dam-breaking bombs that skipped along the water to submerged mines that exploded when they magnetically sensed the presence of a ship's hull, the war was as much a scientific struggle as anything else.

World War II convinced everyone, including skeptical military leaders, that physics is an essential science. Yet the reach of this subject extends far beyond military applications. The principles of physics affect every part of the world and touch on all aspects of people's lives. Hurricanes, lightning, automobile engines, eyeglasses, skyscrapers, footballs, and even the way people walk and run must follow the dictates of scientific laws.

The relevance of physics in everyday life has often been overshadowed by topics such as nuclear weapons or the latest theories of how the universe began. Physics in Our World is a set of volumes that aims to explore the whole spectrum of applications, describing how physics influences technology and society, as well as helping people understand the nature and behavior of the universe and all its many interacting parts. The set covers the major branches of physics and includes the following titles:

- *Force and Motion*
- *Electricity and Magnetism*

- *Time and Thermodynamics*
- *Light and Optics*
- *Atoms and Materials*
- *Particles and the Universe*

Each volume explains the basic concepts of the subject and then discusses a variety of applications in which these concepts apply. Although physics is a mathematical subject, the focus of these books is on the ideas rather than the mathematics. Only simple equations are included. The reader does not need any special knowledge of mathematics, although an understanding of elementary algebra would be helpful in a few cases. The number of possible topics for each volume is practically limitless, but there is only room for a sample; regrettably, interesting applications had to be omitted. But each volume in the set explores a wide range of material, and all volumes contain a further reading and Web sites section that lists a selection of books and Web sites for continued exploration. This selection is also only a sample, offering suggestions of the many exploration opportunities available.

I was once at a conference in which a young student asked a group of professors whether he needed the latest edition of a physics textbook. One professor replied no, because the principles of physics "have not changed in years." This is true for the most part, but it is a testament to the power of physics. Another testament to physics is the astounding number of applications relying on these principles—and these applications continue to expand and change at an exceptionally rapid pace. Steam engines have yielded to the powerful internal combustion engines of race cars and fighter jets, and telephone wires are in the process of yielding to fiber optics, satellite communication, and cell phones. The goal of these books is to encourage the reader to see the relevance of physics in all directions and in every endeavor, at the present time as well as in the past and in the years to come.

 Acknowledgments 

THANKS GO TO my teachers, many of whom did their best to put up with me and my undisciplined ways. Special thanks go to Drs. George Gerstein, Larry Palmer, and Stanley Schmidt for helping me find my way when I got lost. I also much appreciate the contributions of Jodie Rhodes, who helped launch this project; executive editor Frank K. Darmstadt and the editorial and production teams who pushed it along; and the many scientists, educators, and writers who provided some of their time and insight. Thanks most of all go to Elizabeth Kirkland, a super mom with extraordinary powers and a gift for using them wisely.

INTRODUCTION

AS A STUDENT forced to flee Cambridge University during an epidemic in 1665–66, Isaac Newton—later knighted, becoming Sir Isaac—found a lot of time to do experiments. He put this time to good use, discovering the basis for many of the laws of physics he would go on to publish a few decades later. Newton's equations accurately described acceleration and motion, and his universal law of *gravitation* explained in a concise and mathematical way gravity on Earth as well as in the solar system.

The physics of Newton dominated physics for more than 200 years. In Newton's viewpoint, forces caused changes in motion, which could be precisely determined and calculated, and concepts such as space and time were absolute, the same for everyone. Physicists continued to accept this point of view until, in the 20th century, exceptions began to appear. With improved instruments and more imaginative theories, people began to probe objects and events that were not encountered in everyday life—tiny particles inside an *atom,* immense objects such as the entire universe, and small or large objects moving at exceptionally fast speeds. Laws described by Newton failed to hold true in many cases. New laws, and occasionally entirely new concepts, were needed. The new laws reduce to the old laws in familiar situations but increase their scope and accuracy.

Particles and the Universe documents the phenomena in which Newton's physics failed and explains "modern" physics that formed the basis for a new set of laws. One thing that did not change was the scientific method—observations lead to theories, which must be tested for accuracy. Each chapter of *Particles and the Universe* delves into the observations, theories, and tests of a particular topic:

- nuclear physics
- quantum mechanics
- particle physics
- relativity
- cosmology, the study of the universe

Nuclear physics investigates the properties and behavior of the central portion, or *nucleus,* of the atom. This branch of physics has had perhaps the biggest impact on the world in the 20th century because it evolved into knowledge that helped build the most destructive weapons people have ever known. The atomic bombs that ended World War II in 1945, and the weapons race that followed, changed the course of history. But applications of nuclear physics have also provided enormous *energy* for peaceful purposes, generating about 16 percent of the world's electricity.

The strange behavior of tiny particles such as the components of an atom required physicists to revise their theories, as well as the way that those theories are understood and applied. *Quantum mechanics* supplies the equations to describe the motion and properties of particles, but its measurements have peculiar features. Properties of objects tend to have a discrete nature—their values increase by specific amounts, like the integers (. . .–2, –1, 0, 1, 2,. . .) rather than being continuous, like the real number line, in which the value can be any number. Calculations in quantum mechanics also introduce an amount of uncertainty that can never disappear. Physicists dealt with uncertainty before quantum mechanics, but it was due to a lack of knowledge, not due to the nature of physics itself, as it is in the newer theory.

To probe the nature of matter even further, physicists have built gigantic accelerators capable of hurling particles down a pathway at nearly the speed of light. Crashes between high-speed particles have enough energy to tear them apart or to create entirely new particles, and hundreds of different particles exist. Particle physics is the branch of physics devoted to classifying these particles, identifying their properties, and explaining the forces they exert on each other as they interact.

Extremely fast speeds, such as those achieved by particle accelerators, were another phenomenon requiring a fresh perspective in physics. A few decades before huge accelerators were built, Albert Einstein, one of the greatest physicists of all time, concerned himself with the laws of physics as they would appear to observers in motion. Einstein believed physics should be the same for all observers, and his *special theory of relativity,* published in 1905, generated strange but accurate predictions of slowly moving clocks and shrinking lengths. The *general theory of relativity,* proposed a decade later, involved gravitation and had its own astonishing consequences, such as the discovery of objects in space so dense that not even light can escape them. Einstein's theories have survived every test so far.

The special and general theories of relativity are also important tools in the study of the universe. These theories help astronomers understand the observations made with telescopes and other instruments, which reveal a host of spectacular objects and events. One of the most fascinating phenomena is the expansion of the universe itself, a prediction of the general theory of relativity even Einstein refused to believe at first.

All chapters include a description of the profound changes caused by the new discoveries, along with applications such as earth-shattering weapons, machines to image the activity of a human brain, and precise satellite navigation systems. The rise of 20th-century physics altered the landscape of science, producing new ideas and theories that dramatically advanced scientific knowledge in previously unexplored realms of the universe.

1

Nuclear Physics:

Radioactivity, Weapons, and Reactors

EARLY IN THE 20th century a young chemist, George de Hevesy (1885–1966), ate meals prepared by his landlady that he suspected were leftovers. He decided to test his suspicion one day by injecting a small, harmless amount of a chemical into the food remaining on his plate after he finished eating. The chemical was weakly *radioactive:* The atoms emitted *radiation* that could be detected by a sensitive instrument. Later de Hevesy's landlady served him a radioactive meal—it contained some of the food he had left on his plate earlier.

A pioneer of the widely used technique of radioactive tracing, de Hevesy and others led the way in the application of the modern ideas of nuclear physics. The physics of the atom's nucleus has been responsible for a wide variety of earth-shattering developments: *radioactive dating* of ancient materials, nuclear energy reactors that produce huge amounts of electricity, and the most shattering development of all, nuclear bombs. This chapter describes these and other important inventions, along with the underlying principles of nuclear physics.

The nucleus, the object that did so much to shape 20th-century physics, consists of tiny particles called *protons* and *neutrons* and is so small that 100 billion would fit inside the diameter of a human hair.

The Nucleus of an Atom

French physicist Antoine-Henri Becquerel (1852–1908) unknowingly began the era of nuclear physics in 1896 when he discovered radiation coming from pitchblende, a mineral containing uranium. The more uranium in pitchblende, the more radiation it emitted, so the element uranium was important. Polish physicist Marie Curie (1867–1934) coined the term *radioactivity* for these radiation emissions, and, along with her husband, Pierre Curie (1859–1906), discovered an even stronger source of radiation, the element radium. (Radium is such a strong source of radiation that, unlike uranium, pure radium glows in the dark.) Becquerel and Marie and Pierre Curie shared the 1903 Nobel Prize in physics for their work.

Radioactivity is the emission of radiation that accompanies a transformation, called *radioactive decay,* of an atom's nucleus. But it was not until 1911 that New Zealand/British physicist Ernest Rutherford (1871–1937) discovered the atomic nucleus. In a set of experiments that was to have a profound effect on the world, Rutherford directed a beam of particles called *alpha particles* at a thin piece of gold foil. At the time of this experiment scientists did not know what an atom was made of, but many people accepted a theory developed by British physicist Sir Joseph John Thomson (1856–1940). Thomson, a brilliant scientist who had identified the negatively charged atomic particle called the *electron* in 1896, proposed that atoms—which in their normal state are electrically uncharged—were made of clouds of a positively charged substance embedded with electrons. Rutherford decided to probe the structure of atoms by measuring how the atoms of gold scattered the alpha particles. Since alpha particles are positively charged, Rutherford expected that some of these particles would be slightly deflected as they passed through the thin layer of atoms in the sheet of gold. But to Rutherford's surprise, some of the alpha particles bounced off the gold atoms and went straight backward!

The correct interpretation of Rutherford's experiment is that atoms contain within their interior a small nucleus of positive charge. Most of the alpha particles passed through the atoms with

Modern instruments also employ alpha particles. This device, attached to the Mars Exploration Rover *Spirit,* examined the Martian surface by bouncing alpha particles and other emissions off the soil and analyzing the reflections. *(NASA-JPL)*

slight deflections, as expected, but the small number of particles that were reflected straight back had hit a hard, concentrated object. A new picture of the atom emerged: Atoms consist of a small, positively charged nucleus surrounded by a swarm of negatively charged electrons. The positive and negative charges offset each other, giving atoms their electrical neutrality.

A few years later, in 1919, Rutherford identified the positive charges in the nucleus. He named them protons, from the Greek word *protos,* meaning *first.* Protons are 1,836 times more massive than electrons. But later discoveries suggested that the atomic nucleus contained other particles besides the proton, and in 1932 British physicist Sir James Chadwick (1891–1974) discovered the neutron and determined its *mass.* The neutron is slightly heavier than the proton, with about 1,840 times the mass of an electron. The neutron, as its name suggests, is electrically neutral. For his discovery Chadwick was promptly awarded the Nobel Prize in physics in 1935. (Rutherford won the Nobel Prize in chemistry in 1908 for his early work on radioactivity.)

The basic picture of an atom was complete. Protons and neutrons reside in the nucleus, surrounded by a swarm of electrons, as shown in the figure. Electrons and protons have an electrical charge of equal magnitude but of opposite sign, and the normal state of an atom consists of an equal number of electrons and protons. Neutrons add mass to the nucleus but not electrical charge. The number of protons in the nucleus is the *atomic number;* each chemical element has a different atomic number. For example, an atom of the element carbon has an atomic number of 6 (six protons), and iron atoms have an atomic number of 26. The nucleus also contains some quantity of neutrons, but this quantity can vary—atoms of the same element may have a different number of neutrons. Carbon atoms, for example, always have six protons but can have six, seven, or eight neutrons in their nucleus. The mass number equals the number of *nucleons* (protons and neutrons). Atoms of the same element but with a different number of neutrons are called *isotopes.* The isotope of carbon with six protons and six neutrons has a mass number of 12 (designated as carbon 12), and the isotope with six protons and eight neutrons has a mass number of 14 (carbon 14).

The simple picture of an atom remains an important part of physics. Recent discoveries have complicated the picture, as described later in this book, but the basic description of the atom has remained essentially correct.

A puzzling aspect of the atomic nucleus is the concentration of positively charged protons. Opposite charges attract—protons and electrons attract one another—but like charges repel, so how can the protons exist packed tightly together in the nucleus? There is clearly another force acting on the protons, and physicists call it the *strong nuclear force* or often just the *strong force.* This force is strong (hence the name), and it binds both protons and neutrons together in the nucleus. But the particles must be close to one

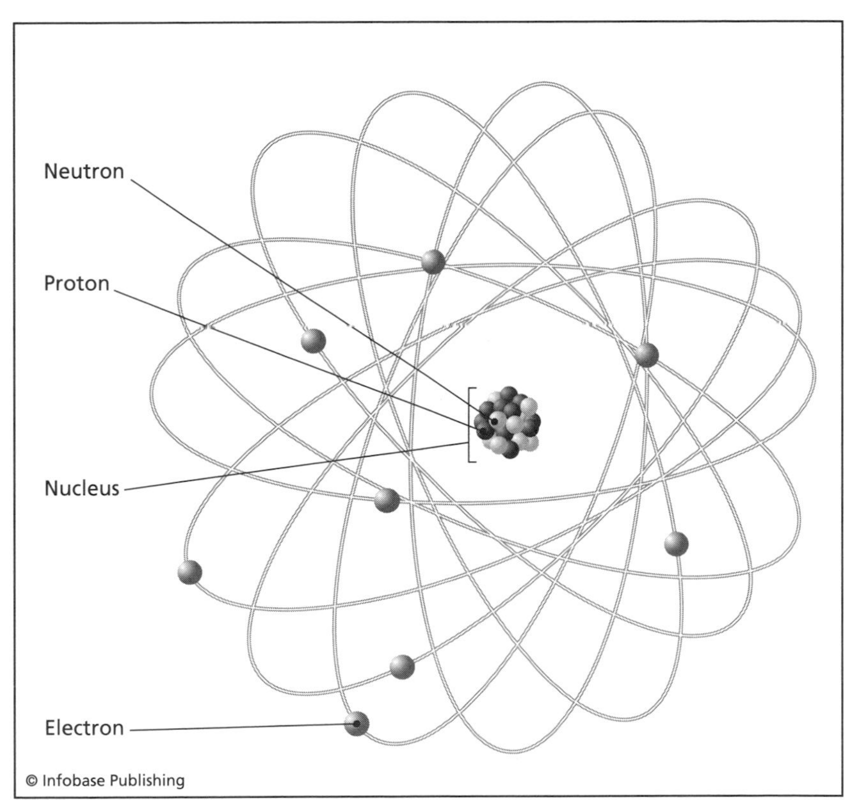

Electrons of the atom swarm around the nucleus. The atom's nucleus consists of neutrons and protons packed tightly together.

another because the strong force weakens rapidly with distance. Protons normally repel one another because of their charge, but when they are next to one another the strong force overcomes the electrical repulsion. This means that free protons usually will not come together automatically to form a nucleus, but under the right circumstances—if they crash into each other at high speed, for instance—the strong force will bind them together. (There is also a *weak nuclear force,* often called the *weak force,* which is another short-range nuclear force. The weak force is responsible for many of the radioactive processes described later, such as the production of *beta particles.* The strong force is about a billion times stronger than the weak force.)

Radioactivity arises when the nucleus of an atom decays. In some atoms the nucleus is unstable; like a stick balanced on its edge—which has a tendency to fall—an unstable nucleus is subject to change. The change involves the emission of radiation, as described in detail in the sidebar titled "Radioactivity." This process also often causes a transformation in one or more of the protons or neutrons in the nucleus, resulting in a different element if the number of protons changes. For example, carbon 14 is radioactive and decays into nitrogen 14; uranium 238 decays into thorium 234. For many years ancient scientists called alchemists unsuccessfully sought methods to transform elements such as lead into gold, never realizing that some elements transform themselves naturally, thanks to radioactivity.

Because of the constant decay rates of radioactive isotopes, as explained in the sidebar, radioactivity is like a clock. The clock-like precision of decay permits its use in the dating of materials, a process called radioactive dating. If the original quantity of radioactive material is known, its age can be determined by measuring how many radioactive atoms remain and then applying the half-life rule, as described in the sidebar on pages 8–9. For instance, a sample that has decayed to one-eighth of its original amount is three half-lives old.

Sometimes the original quantity of radioactive material can only be crudely estimated. But for a number of isotopes that are found on Earth, such as uranium, thorium, and others, the ratios of these isotopes and their known decay rates provide an accurate measurement of the age of the rocks in which they are embedded. These measurements reveal Earth's age, and physicists are confident of their accuracy because they have tested a number of different isotopes and ratios and the results agree that Earth is about 4.5 billion years old. Samples from meteorites also give this same age, which means that the rest of the solar system formed at about the same time.

Another accurate dating method uses carbon 14. This is the method to date organic, or life-related, material, because organic material contains carbon. Called radiocarbon dating, the procedure works because all organisms are constantly taking in material that contains carbon. Animals eat food containing plenty of carbon in the carbohydrates, proteins, and fats, and plants use carbon dioxide and sunlight to make carbohydrates. About one carbon atom in a trillion on Earth is carbon 14, which is radioactive with a half-life of 5,700 years. (Most carbon atoms are stable isotopes such as carbon 12.) While a plant or animal remains alive, its number of carbon 14 atoms remains constant, since the organism continually consumes new carbon. But when it dies the plant or animal no longer takes in carbon, so the carbon 14 in the body decays and is not replaced. By measuring the amount of carbon 14 remaining, scientists can determine how long ago the organism was alive. Radiocarbon dating is exceptionally accurate with objects a few thousand years old, but after about 50,000 years there is so little carbon 14 left that measurements become too difficult.

A well-known example of radiocarbon dating in action is "Iceman." A group of hikers climbing the mountains between Austria and Italy in 1991 found a mummy—the remains of a man who had died and whose body was preserved by the snow and ice. Radiocarbon dating placed the time of death about 5,300 years ago (which, along with the environment, made it truly a "cold case"). Iceman, also known as Ötzi, is one of the oldest and best-preserved mummies ever discovered.

Radioactivity has other uses besides dating objects. One of the most common applications is present in nearly every building—smoke detectors. These devices sense the presence of smoke in the air and alarm the occupants of the fire. One type of smoke detector, called a photoelectric smoke detector, shines a beam of light through a sample of air; this works somewhat like a ray of sunshine that comes through a window and highlights the dust and smoke particles floating around in the room. But these smoke detectors overlook the smallest particles, which are not big enough to reflect much light. A more effective type of smoke detector uses a tiny amount of radioactivity to ionize the air. The radioactive isotope in these ionization smoke detectors is generally americium 241. As mentioned in the sidebar "Radioactivity," radiation rips the electrons from atoms, forming electrically charged *ions* that allow the smoke detector to pass an electric current through a sample

Radioactivity

Physicists who studied radioactivity in the early 20th century discovered that radioactive emissions consist of different types of radiation. Rutherford noticed two types that could be distinguished based on how easily they were absorbed. He named the first type—which could be stopped by a piece of paper—alpha particles, and the second type, which were not absorbed as easily and could penetrate further into objects, beta particles. Alpha (α) and beta (β) are the first two letters in the Greek alphabet (a word that is also derived from the first two Greek letters). French physicist Paul Villard (1860–1934) discovered another type of radiation that was more energetic than alpha and beta particles. Continuing with the Greek alphabet, this radiation is a *gamma ray,* named after the third Greek letter, gamma (γ).

Alpha particles are positively charged, consisting of two protons and two neutrons bound together. This is the same as a nucleus of the element helium, and sometimes physicists refer to alpha particles as such. Uranium 238, for example, decays into thorium 234 by emitting an alpha particle. Beta particles are negative charged—they are electrons. An example of decay involving beta particles is carbon 14 into nitrogen 14. (A neutron in carbon 14 decays into a proton and emits an electron, the beta particle, which escapes because of its high energy. The nucleus contains the same number of nucleons after the decay but has one more proton and one less neutron than before.) Gamma rays are *electromagnetic waves* (also known as *electromagnetic radiation*), like light, except of a much higher frequency. The energy of electromagnetic radiation depends on its frequency—higher frequencies have higher energies. Gamma rays have frequencies of about 10^{20} *hertz* (cycles per second) or more, two million times higher than visible light, so they possess enormous amounts of energy. Gamma-ray decay occurs when a radioactive nucleus is in an excited or energetic state. The emission of a gamma ray does not change the nucleus except for lowering its energy.

About 3,500 isotopes exist and many are radioactive, including all nuclei (plural of *nucleus*) with an atomic number greater than 83—all these heavy isotopes are unstable and decay. The majority of the 3,500 isotopes are not present on Earth in any significant amount and have been made or found in laboratory experiments, but the world does contain plenty of radioactive isotopes from elements such as carbon, uranium, and others.

Two factors determine the quantity of radioactivity in a material. One factor is the number of radioactive atoms; as with pitchblende, more radioactive atoms emit more radiation. The second factor is the specific type of radioactive isotopes present in the material. Some isotopes, such as potassium 40, are unstable and decay, but they take their time. Like a stick that is almost but not quite balanced on its edge, these nuclei are relatively more stable than other radioactive isotopes and do not decay quickly. Other isotopes, such as bismuth 211, are more unstable and decay promptly. The quick-to-decay radioactive isotopes produce a lot of radiation per unit time.

An important characteristic of a radioactive isotope's decay rate is that it is constant. The rate is not affected by location, condition, or anything else. Half of the original quantity of material decays in a certain interval of time, then half of the remainder decays in an identical interval, and so on. Physicists use the term *half-life* to describe the decay rate of isotopes; the half-life is the amount of time for one-half of the sample to decay. The half-life of potassium 40, for example, is about 1.25 billion years, but the half-life of bismuth 211 is 2.15 minutes. This means that it takes 1.25 billion years for half of the atoms in a piece of potassium 40 to decay, another 1.25 billion years for half of the remainder to decay (which would leave a quarter of the original material, half of a half), and another 1.25 billion years for half of that to decay (leaving one-eighth the original), and so on. After $3 \times 1.25 = 3.75$ billion years, one-eighth of the potassium 40 still exists and seven-eighths has decayed. Bismuth 211 would be down to one-eighth the original in 6.45 minutes.

Radioactivity is dangerous, though everyone receives small doses throughout their lives. The human body even contains a tiny amount of radioactive isotopes, such as carbon 14 and potassium 40. In large doses the radiation from radioactive isotopes damages molecules such as *deoxyribonucleic acid* (DNA) by *ionization*—the radiation has enough energy to tear electrons away from atoms, causing them to become electrically charged or ionized. (Gamma rays are particularly dangerous forms of ionizing radiation because of their high energies.) In DNA, which carries an organism's genetic information, the damage can cause genetic changes called mutations that are usually harmful. Instruments such as Geiger counters and dosimeters detect and measure radioactivity, in many cases by detecting ionization. People who work in places subject to radioactivity must be monitored in order to limit their radiation exposure.

of air. The current is usually steady. But ions stick to any smoke particles present in the air instead of carrying the current, so when the current diminishes, the detector sounds an alarm.

These applications of radioactivity display its usefulness, as well as the ingenuity of engineers who develop the applications. But perhaps the most important aspect of radioactivity was that it gave physicists their initial glimpse of the nucleus of the atom, and the vast quantity of energy it possesses.

The Atomic Bomb

When World War II began in 1939, physicists had been probing the nucleus for a few decades and understood its basic principles. Radioactivity showed that nuclei had plenty of energy, and people wondered if it was possible to harness that energy.

Around this time in the 1930s, German physicist Hans Bethe (1906–2005) solved the mystery of the Sun's prodigious output by recognizing that stars are fueled by nuclear reactions—processes by which atomic nuclei are transformed and release huge amounts of energy in the form of fast-moving particles and electromagnetic radiation, including visible light. Before the discoveries of nuclear physicists, no one knew the true source of the Sun's energy; if the energy was from burning coal or oil (substances familiar to 19th- and early-20th-century scientists), the Sun would shine for only a few million years before using all its fuel. Bethe proposed that the Sun could have been lighting up the sky for much longer if it were powered by nuclear reactions, and he proved to be correct: The Sun has been around for 4.5 billion years and will last at least a few billion years more.

Spontaneous decay of nuclei such as radioactivity is one way that the energy of the nucleus is released. Two other ways are *fission*—the splitting of the nucleus into two or more smaller parts—and *fusion,* which occurs when nuclei come together to form a heavier nucleus. Nuclear weaponry uses both of these reactions.

Protons and neutrons in the nucleus are normally held tightly together by the strong force. There is a barrier, created by the strong force, which must be overcome if nucleons are to escape.

Protons are always trying to push one another out of the nucleus because of electrical repulsion, but the electrical force is not powerful enough to do the job alone. Add a little energy, though, and the nucleons can escape. Sometimes one of the nucleons escapes or is transformed spontaneously—without any help—as in the radioactivity discussed in the previous section. What causes these decays are mostly the strange effects arising from quantum mechanics, which will be described in detail later in the book. Radioactive decays are events more likely to happen when a nucleus contains many particles, since more opportunities exist for spontaneous decay. Having more protons in a nucleus means that more of them are trying to shove each other out, which is one of the reasons that large nuclei are unstable.

Fission, however, is a process that usually needs a push to get started. Once it gets that push the reaction may go on to release a frightening quantity of energy. As World War II began, military strategists started to talk about weapons made from atoms and nuclei—the atomic bomb. Motivated in part by fears that Germany would create such a devastating weapon to win the war, the United States initiated a program known as the Manhattan Project to design and build an atomic bomb. One of the most prominent scientists who claimed that such a bomb was possible was German-American physicist Albert Einstein (1879–1955), who emigrated to the United States in 1933 to avoid persecution by Germany's Nazi government (Einstein was a Jew, millions of whom were murdered by the ruthless Nazi regime). Einstein did not work on the bomb itself, but years earlier he had discovered the secret of nuclear energy: the equation $E = mc^2$, where E stands for energy, m is mass, and c is the speed of light. The following sidebar provides more information on this famous equation.

Another important finding that led to the development of the atomic bomb was neutron-induced fission. Italian physicist Enrico Fermi (1901–54) and his colleagues found in 1934 that when neutrons were added to uranium, fission occurred. (At first Fermi and his group did not realize what they had accomplished. The process was only understood a few years later.) As shown in the

$E = mc^2$

Albert Einstein discovered this equation not in the study of atomic nuclei but rather as a part of his theories on light and motion. Einstein was an amazing theorist who performed no experiments except those he could do in his mind, and he seemed capable of grasping even the most difficult and abstract of ideas. Einstein offered a set of novel concepts in 1905 that became known as the special theory of relativity. Einstein believed that the speed of light in a vacuum—186,200 miles/second (300,000 km/s) and denoted c—was the speed limit for all objects in the universe. He theorized that no object could be accelerated up to or past this speed, and he derived an equation that related the energy of an object to its motion. An interesting consequence of this equation arises when the object's velocity, v, was set equal to 0: The result was $E = mc^2$. This energy is the object's "rest energy." (Einstein's relativity theories are discussed in more depth in chapter 4.)

Einstein's ideas proved correct. What the equation $E = mc^2$ says is that energy and mass are interchangeable. This was a strange concept, for although physicists had long realized that energy could be transformed from one form to another—chemical to mechanical, for instance (which is how cars burn fuel and accelerate)—no one had thought much about energy transforming into, or out of, mass. Mass was considered to be the quantity of matter, and according to the great British physicist Sir Isaac Newton (1642–1727), mass was related to an object's inertia

figure, when a neutron hits and sticks to a uranium 235 nucleus, the nucleus becomes highly unstable and fragments into smaller nuclei, such as rubidium and cesium. Other products of the fission process in this case are free neutrons, unattached to any nucleus. These neutrons would seem to be unimportant products, compared to the nuclei, but this is quite false. The neutrons turned out to be the key to creating a *chain reaction* that unlocked the vast energy of the nucleus.

American scientists who developed the atomic bomb during World War II focused their efforts on this fission process. To make an atomic bomb, one needed a fissionable material and the means (resistance to motion). But Einstein showed that mass could be transformed into energy, and the exchange rate is fantastically large. A tiny speck of mass makes a whale of an explosion because c is a huge number and when it is squared, as in c^2, it becomes even bigger.

Nuclear reactions get their energy from $E = mc^2$. Fission, for example, occurs when a nucleus splits into two or more pieces. The sum of the masses of the reaction products do not add up to the mass of the original—the sum is slightly less, for some of the mass vanishes. This missing mass is transformed into energy. The same phenomenon also occurs in chemical reactions, such as the combining of sodium with chlorine to form table salt, sodium chloride. This reaction releases heat and light energy, and the mass of the product, sodium chloride, is slightly less than the mass of the reactants, sodium and chlorine. But the missing mass in the case of the salt reaction is only one part in several billion, too small to be detectable, which is why students of chemistry correctly claim (to the limit of their measurements) that mass is conserved in chemical reactions.

Nuclear reactions transform much more mass on a percentage scale than do chemical ones. Chemical reactions involve electrons, the lightweights of the atomic world; nuclear reactions involve the nucleus of the atom, where 99.9 percent of the mass resides. In the fission of uranium atoms, to be described later, the amount of missing mass is about 0.1 percent. Even that seems a piddling amount, but the energy generated is impressive, thanks to the c^2 in $E = mc^2$.

to induce fission in a lot of nuclei at the same time. Uranium fission converts about 0.1 percent of the mass of the uranium nucleus into energy, and while Einstein's equation guarantees that this is proportionally a great quantity compared to the mass, the energy from one or only a few nuclei generates more of a firecracker than a powerful bomb. This is where Fermi's neutron-induced fission comes into play, since uranium fission produces more neutrons than it consumes. The number of neutrons generated during the process exceeds the number required to set it off (each nucleus that fissions yields one to three neutrons, an average of 2.5 per event). A few neutrons fired at heavy nuclei such as uranium 235

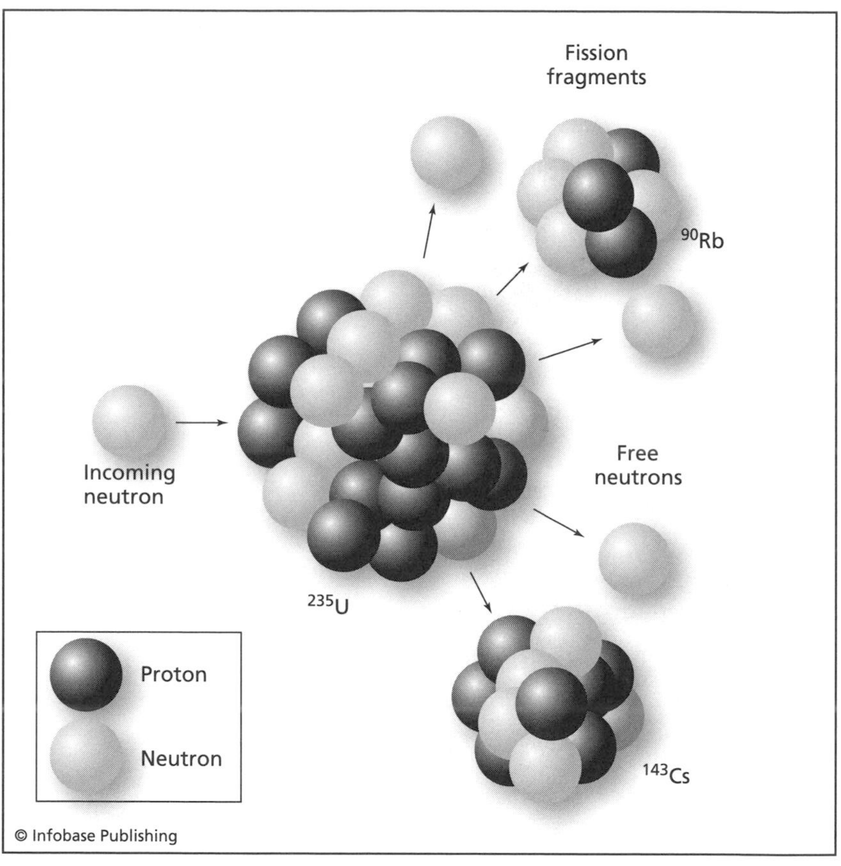

The fission of a uranium nucleus produces two smaller nuclei and one to three neutrons.

and plutonium 239 induce fission in some of the nuclei, and these events produce more neutrons that in turn induce more nuclei to undergo fission. A chain reaction occurs. And, to the delight, and fright, of the physicists working on the Manhattan Project, a devastating bomb was born.

Chain reactions require a minimum amount of material. If only a few nuclei undergo fission, only a few neutrons are generated and the reaction sputters and dies before it can get going. The *critical mass* is the quantity of material sufficient to support a chain reaction. For uranium 235, the critical mass is about 114 pounds (507 N).

The first atomic bomb exploded in a test conducted in the New Mexico desert near Alamogordo on June 16, 1945. This device used plutonium 239 and released energy roughly equivalent to the detonation of more than 20,000 tons of the high-explosive compound trinitrotoluene (TNT). The development of the atomic bomb came too late to be used in the war with Germany (which surrendered on May 7, 1945), but World War II continued because Japan refused to give up. The Americans and their allies were faced with the prospect of invading the heavily fortified Japanese islands, with a loss of potentially millions of lives—soldiers from both sides as well as Japanese citizens. The desire to avoid such an invasion prompted the United States to drop two atomic bombs. The first, on August 6, 1945, fell on the Japanese city of Hiroshima, bursting with the power of about 13,000 tons of TNT. This bomb was made from uranium 235. Another bomb, similar to the one used in the New Mexico test, fell on the Japanese city of Nagasaki on August 9, 1945, after the initial target, the city of Kokura, was obscured by clouds. These bombs killed more than 100,000 people immediately and horribly injured thousands more, many of whom died later of their wounds. Japan surrendered on August 14, 1945.

A 60,000-foot (18,300-m) column of smoke and dust signals the devastation of the atomic bomb dropped on Nagasaki, Japan, in World War II, on August 9, 1945. *(National Archives and Records Administration)*

Although the atomic bombs brought an end to World War II, they also ushered in an age in which militaries had acquired an awesome power. The cold war, beginning in the late 1940s and continuing through the 1980s, was a

The atomic bomb destroyed much of Nagasaki, Japan. *(National Archives and Records Administration)*

time when the strongest countries that emerged from World War II—the United States and the Soviet Union—faced each other in a dangerous standoff. Although the countries had fought on the same side during World War II, different political ideologies—systems of government—made them enemies afterward. Both countries stockpiled a deadly arsenal of nuclear weapons.

Weapons based on fission were not the only bombs in the arsenal. Only a few years after World War II, United States physicists developed a bomb based on fusion. Fusion occurs when small nuclei join to make a larger nucleus, the opposite of fission. For example, isotopes of hydrogen called deuterium (which is hydrogen 2, containing a proton, like normal hydrogen, plus a neutron) can combine to make a helium 4 nucleus. As in fission, the process releases a tremendous amount of energy, and for the same reason ($E = mc^2$). Even more mass, on a percentage basis, is lost than in fission; for example, when deuterium fuses with another hydrogen isotope called tritium (hydrogen 3, one proton and two neutrons), the products are helium 4 and a free neutron, plus the conversion of roughly 0.3 percent of the original mass into energy. These fusion bombs were called "H-bombs" after hydrogen, the fuel for the explosion. (Fusion will be described in more detail in a later section of this chapter.)

Many nuclear weapon tests occurred as the United States, the Soviet Union, and other countries such as France and England developed bombs capable of even more devastation. The most powerful nuclear weapon tested to date was a Soviet bomb in 1961, which exploded with the energy of 50,000,000 tons of TNT, almost 4,000 times the destructiveness of the Hiroshima device. (This bomb, tested in the Arctic Circle, became known as *Tsar Bomba,* "King of the Bombs.") Both sides of the cold war developed missiles of increasing range and sophistication and aimed them at each other. Preventing a trigger-happy release of these terrible weapons was a darkly odd concept known as MAD—Mutual Assured Destruction—by which both sides were guaranteed to be destroyed in any war involving nuclear weaponry.

Today the political tension that caused the cold war has eased due to the disbanding of the Soviet Union. But nuclear weapons remain. As of May 2006, the number of countries that acknowledge possessing nuclear weapons is eight: United States, Russia, United Kingdom, France, People's Republic of China, India, Pakistan, and North Korea. In addition, Israel is suspected of owning nuclear weapons, and Ukraine (formerly part of the Soviet Union) may also harbor nuclear missiles. Many countries have signed the Nuclear Nonproliferation Treaty in the attempt to limit the spread and development of these dangerous weapons, but many people continue to worry that a small country or even a group of determined terrorists will acquire nuclear capability and fail to show restraint. In the hands of foolhardy, militaristic leaders, nuclear weapons have fearful consequences—not just for their awesome destructive power, but also for their enduring radioactivity, known as *fallout.*

Radiation is a consequence of spontaneous radioactivity and induced nuclear reactions. As discussed earlier, radiation from just a few radioactive atoms is not hazardous but in larger doses will kill or injure. Explosions of fusion or fission bombs usually produce a large number of nuclei that are subsequently unstable and decay, with half-lives of a few seconds to thousands of years. These nuclei disperse through the air, drift with wind, attach themselves to small dust particles, and, in general, can travel for miles and contaminate every exposed surface as the radioactive substances finally fall to Earth. Cancer, caused by mutations in DNA damaged with ionizing radiation, increases with exposure to radioactivity.

Such a disaster occurred in 1954 during a nuclear bomb test called BRAVO conducted by the United States at the Bikini Atoll, a small island in the Pacific Ocean. The blast exceeded expectations, turning out to be 1,000 times stronger than the bomb dropped on Hiroshima. Residents of neighboring islands suffered from severe radiation exposure, causing vomiting, diarrhea, and burns. Years later many of the children developed cancer. Although the United States admitted the error and provided health care and compensation, the disaster took a horrible toll.

A nuclear weapon disaster can even affect people on a global scale. Detonation of a large number of bombs simultaneously, such as during a world war, might throw up enough dust and debris to block out sunlight for months or even years. In addition to the fallout, which would kill millions, Earth would experience reduced temperatures because of the loss of solar energy. Similar events in the history of the planet have been caused by large meteorite strikes and are associated with the extinction of numerous species.

Nuclear Energy

The nightmares associated with nuclear weapons disillusioned many people on the concept of nuclear energy. Yet nuclear power provides approximately 16 percent of the world's electricity supply today, according to the International Atomic Energy Agency.

Nuclear reactors are power production facilities. The idea is the same as any other electrical power station, to generate electricity by transforming the energy extracted from a potential energy source. The potential energy may come from chemical sources,

Argonne National Laboratory in Illinois built the first nuclear reactor in 1942. This photograph shows the nuclear material used as fuel, uranium oxide pellets in a graphite (carbon) block. *(Argonne National Laboratory, courtesy AIP Emilio Segrè Visual Archives)*

such as burning coal, oil, or gas, or it may come from waterfall, wind, or sunlight. The source of energy in nuclear reactors is from the nucleus of the atom, the same place and the same concept that can produce horrific explosions. But reactors apply Einstein's equation ($E = mc^2$) in a slow, controlled manner.

The process of extracting energy in a controlled fashion is exactly what organisms do when they digest food. Burning the carbohydrates and fats in food generates a certain amount of energy. When the combustion takes place in a furnace it happens quickly, releasing heat and light energy. The digestive system of an organism extracts this same energy—which is chemical potential energy, stored in the chemical bonds of the food molecules—but the process is slow and controlled. The body dismantles food molecules piece by piece, and the liberated energy is acquired by the energy-storing molecules of the organism, usually adenosine triphosphate (ATP). Energy in the chemical bonds of ATP molecules powers much of the activity of cells in all organisms on this planet.

Nuclear reactors generate electricity by the slow and steady fission of unstable radioactive isotopes. Most reactors use uranium.

Fusion of nuclei would also yield energy and in many ways would be more desirable than fission, particularly since the fuel—for example, deuterium in water—is cheap and plentiful. But while bombs have been made from the fusion process, there are presently no operational fusion reactors because fusion is exceptionally difficult to control. Physicists and engineers are working on this problem, as discussed in detail in a later section of this chapter.

The purpose of a fission reactor is to create a self-sustained nuclear reaction. The process must continually supply energy from the fission of nuclei, induced by neutrons, like the chain reaction that leads to the explosion in the atomic bomb. But in a reactor the fission must be allowed to produce only enough neutrons to keep the process going at a steady rate. To achieve this, the number of neutrons generated by fission must equal the number of neutrons absorbed by the nuclei undergoing fission. If each uranium atom that undergoes fission produces, on average, one neutron to be absorbed by another uranium atom that subsequently splits, the process will be self-sustaining but will not lead to a runaway explosion—there are no excess neutrons. A process that produces fewer than one neutron per fission event will eventually die out, so the number of neutrons required is one, no more and no less.

But uranium fission is a natural event that need not cooperate with human engineering goals. As mentioned earlier, uranium 235 fission yields about 2.5 neutrons per event. A reactor therefore needs a mechanism to prevent some of these neutrons from triggering another fission event. Accomplishing this is the role of control rods, inserted into the reactor core—the area in which the reaction proceeds. The figure below illustrates a reactor core. The control rods are composed of a material such as cadmium or boron that absorbs neutrons but does not undergo fission. Adjusting the number and position of control rods in the core gives the operator a mechanism to control the activity of the free neutrons. When the activity is too low, control rods can be raised or removed so that fewer neutrons are absorbed and the reaction rate increases; if the activity becomes too high, the opposite procedure decreases the rate.

Another issue in reactor cores is the requirement to slow down the neutrons. Fast-moving neutrons are not as easily absorbed

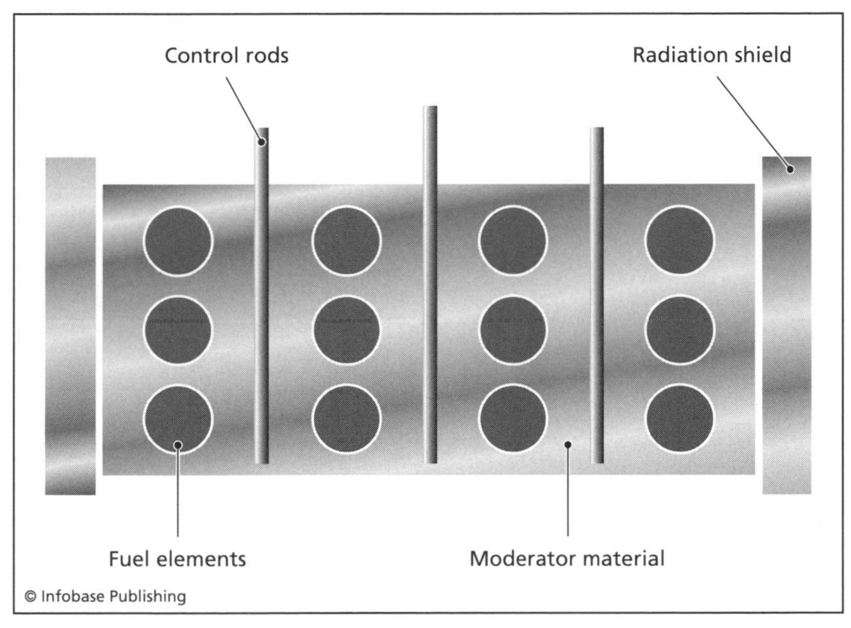

A slice through a reactor core shows the fuel elements that undergo fission, the moderator material, movable control rods, and shields to absorb stray radioactive emissions.

by the isotope of uranium that is used as fuel (uranium 235), but the neutrons generated in the fission events tend to be energetic. The job of slowing down the neutrons belongs to the "moderator" material, which in many reactors is water. Neutrons bounce off water molecules and are not absorbed but instead are slowed down to a speed that encourages their capture by a uranium 235 nucleus.

An additional reason for slowing down the neutrons is that uranium 238, a different isotope of uranium, absorbs a lot of fast-moving neutrons but rarely undergoes fission. Uranium 238 is a "dead weight" as far as nuclear fuel is concerned, yet it must be taken into consideration because natural uranium ore is about 99 percent uranium 238 and less than one percent uranium 235. Unlike different chemical elements, isotopes of the same element can be difficult to separate, and this is certainly true of the isotopes of uranium. Uranium must be enriched in uranium 235 to about 3 or 4 percent in order to be useful as fuel for most reactors, and even this is a difficult refining job involving a separation process based on the tiny differences in the mass of the isotopes.

The energy obtained from fission comes from the reduction in mass of the products ($E = mc^2$), but this energy is not in the form of electricity; rather it is heat and radiation. In a reactor, the heat energy turns water into steam. The high pressure of steam turns a turbine, which produces the relative motion required to generate electricity by induction. According to the International Atomic Energy Agency, there are 441 nuclear reactors in 32 countries operating or under construction worldwide as of 2002. Of these, 104 are in the United States and account for approximately 20 percent of the electricity generated in this country.

Fission produces a great deal of energy for a small amount of fuel—this is its advantage. The disadvantage is that it can be dangerous. But a nuclear explosion, like that of a bomb, is not possible in reactors. Such explosions are difficult to achieve and require crunching together a large mass of fissionable material. Instead, the most serious danger of reactors is that they may suffer from heat damage and allow radioactive material to escape. Allowing too many neutrons to wander around in the core will overheat the reactor, possibly resulting in melting the core itself. Temperatures can soar to thousands of degrees, high enough to melt anything. Nuclear reactors are shielded with radiation-blocking material such as thick slabs of lead, but in a catastrophic meltdown, radioactive nuclei may be released. Control rods are supposed to prevent overheating, but they take time to work. Backup systems exist to shut down the core quickly in the case of an emergency.

Considering the number of reactors in the world and the length of time they have been in operation, the accident rate is quite low. The first commercial reactor appeared in 1954 in Russia, and the first commercial reactor to begin operation in the United States was the reactor in Shippingport, Pennsylvania, in 1957. Since that time, the worst accident in the United States happened at the reactor on Three Mile Island in 1979. This facility, which sits in the middle of the Susquehanna River in Pennsylvania, suffered a partial meltdown when a water-circulating pump failed. The loss of cooling water caused the core to overheat, even though the reactor was shut down with control rods, because radioactivity created by the previous fission events continued to emit energy. A tiny amount of radioactivity escaped, although not enough to have a harmful effect. But it did cause some frightening moments for people in the immediate vicinity, including the residents of Harrisburg, the capital of Pennsylvania.

By far the most serious nuclear accident occurred on April 26, 1986, at the Chernobyl nuclear reactor in the Ukraine (then part of the Soviet Union). The incident began when operators, while conducting a test of the cooling system, withdrew a large number of control rods. When the cooling system was turned off, the control rods should have been automatically reinserted to shut down the reaction, but the mechanism had been overridden. The core rapidly overheated, and the operators then tried to shut it down by moving the control rods manually, but it was too

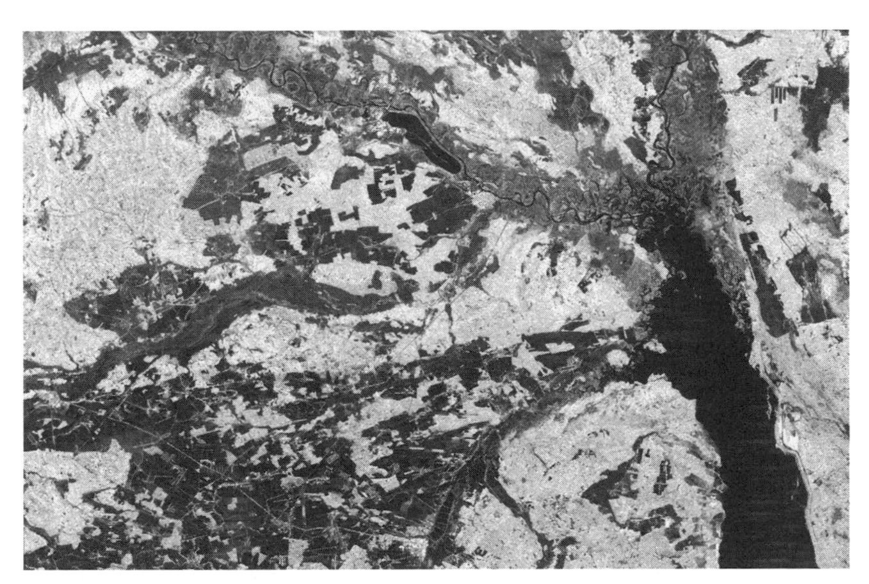

A radar image of Chernobyl, taken from the space shuttle *Endeavour* in 1994, shows the 7.4-mile (12-km) cooling pond for the Chernobyl nuclear reactor as a long, dark shape about halfway between the center and top of the image. To the left of the pond are the reactor and buildings, and to the lower right of the pond, along the Pripyat River, is the city of Chernobyl. The Soviet Union evacuated more than 100,000 people after the accident on April 26, 1986. *(NASA-JPL)*

late. The fission rate had soared and the core melted. Steam and chemical explosions occurred that sent a significant quantity of radioactive nuclei into the atmosphere. More than 30 people died in the initial event and many more were hospitalized. The radioactive material scattered across much of Scandinavia and eastern Europe, forcing hundreds of thousands of people to flee their homes. The aftermath will probably continue to create problems by inflicting an untold number of people with cancers and other diseases due to radiation overexposure.

Nuclear power has obvious concerns. Not only do accidents have the potential to release harmful amounts of radioactivity, but the nuclear reactor fuel also continues to be radioactive for centuries even after it is too far spent to be usable in power generation. The radioactive "waste" must be sealed and stored for many years in underground caves, and the possibility of container corrosion and eventual release makes many people uneasy about having a dump site in their area.

Although nuclear reactor construction in the United States has come to a halt since the early 1980s, other countries consider nuclear power the best alternative to fuels such as oil, coal, and gas. These hydrocarbons are common fuels, but when burned they emit greenhouse gases that are suspected of causing global warming. Nuclear power produces extremely few greenhouse gases. In 2002, France generated 78 percent of its electricity from nuclear energy, Belgium 57 percent, Sweden 46 percent, and even the Ukraine, the site of the worst accident, produced 45 percent of its electricity from its reactors. Although the world's uranium 235 supply is limited and will not last many more decades, while it is available there are people in the world who believe the risk-reward ratio of nuclear energy is favorable.

Nuclear Medicine

Radiation from nuclear reactor accidents is dangerous because of its ionizing power. As mentioned before, ionization strips electrons from atoms and molecules, creating ions. These ions are strongly reactive—new compounds are formed, and old ones are broken down. Living organisms are particularly susceptible because ionization damages or destroys important biological molecules such as DNA. If a cell receives too much radiation it will die.

But ionizing radiation does have a benefit. There are occasions in medicine when a physician wants certain cells in the patient to die. One such occasion is in the treatment of cancer.

Cancer is a disease that comes in many different forms and strikes many different organs and systems, but the one common factor among them is that all cancers are uncontrolled growths. Growth is essential for life, particularly in young organisms that have yet to reach full size, but also for adults, as cells and tissues throughout the body are constantly being replaced. Red blood cells, for example, have a limited lifetime, and about a billion of these cells die each day in an average person. The red blood cells are replenished by a process of cell division—cells grow and split in two, making copies of themselves. Cell division is the normal way that organisms grow, maintain, and repair tissues and organs.

All these cell divisions must be controlled and coordinated so that the proper number of cells is produced when and where they are needed. A cancer develops when a group of cells, or perhaps only a single cell, begins to divide again and again. Quite a few internal mechanisms exist in cells to prevent this from happening too often, and most cells that undesirably begin to replicate themselves will die. It is a rare event for a cell to escape these restraints and start uncontrolled replication. In most of these cases, such a cell has suffered a number of mutations in its genes, all of which combine in an improbable event to allow the cell to bypass the bounds on its growth. An unfortunate aspect of the mutations is that they are inherited, just like other genetic factors. Cell division requires duplicating the cell's DNA, and each of the two daughter cells receives a full share. For a cancerous cell this means that the mutations are passed along to the daughter cells, and they, also, divide uncontrollably. This is how a mass of unneeded tissue such as a tumor forms.

Cancer is the second leading cause of death in the United States. Decades ago cancer was so dreaded that its name was almost unspeakable. Today cancer is still dreaded but it is becoming more survivable, thanks to improvements in treatment. Treatments involve killing or removing the cancer cells with as little damage as possible to healthy tissue. One way to kill a cell is with radiation.

About half of all cancer patients receive some form of radiation therapy, either alone or in combination with other medical procedures. Most often the radiation is delivered by a beam of radiation coming from a source of high-energy particles or electromagnetic radiation. In some cases the beam is applied to the surface of the body, but more commonly the treatment is needed deeper inside. Radiation therapy is useful in the treatment of most solid tumors (masses of abnormal growths), including common cancers of the brain, stomach, lung, and other organs. The amount of necessary radiation depends on a number of factors such as the type of cancer, where it is located, and how far the disease has progressed. Finding cancer at an early stage is generally best; certain types of cancer spread throughout the body and are enormously difficult to treat in advanced stages.

Some of the radiation used in these procedures is not from radioactive sources. X-rays and gamma rays generated by particle accelerators are commonly employed, as are the particles themselves, such as electrons, protons, and neutrons. (Particle accelerators will be discussed in chapter 3 of this volume.) But many treatments come from cobalt 60, a radioactive isotope with a half-life of slightly more than five years. Cobalt 60 is an excellent source of gamma rays and is produced in significant amounts by exposing natural (nonradioactive) cobalt to a barrage of neutrons. Although cobalt 60 therapy produces radioactive waste (the isotope remains radioactive beyond its useful therapeutic application, like other nuclear "fuel"), it is a relatively cheap alternative for hospitals and clinics without access to accelerators.

The goal of radiation therapy is to kill the cancer cells without affecting healthy cells. But directing a beam of radiation, no matter its source and nature, cannot be as precise as physicians would like. Some healthy tissue is damaged, although most cells have mechanisms by which the damage done by radiation can be repaired if it is not too great. Cancer cells, fortunately for the patient, tend to lack some or all of these mechanisms so they succumb more readily. Yet radiation therapy always involves a certain amount of "side-effect" or "collateral" damage. The unintentional damage in many cases occurs to tissues composed of cells that divide frequently (though unlike cancers, the growth and division remains under control). Common examples include hair follicles and cells in the blood, skin, and intestine. This explains the most common side effects of radiation therapy, hair loss and nausea.

The term *nuclear medicine* is sometimes limited to the injection of radioactive substances into the body, as opposed to the use of external sources of radioactivity as in many forms of radiation therapy. Injection is certainly the most direct use of nuclear activity in medicine, though as described above, radioactive isotopes can perform a large number of different functions in medical settings even outside the body.

Once inside the body, radioactive isotopes are useful for diagnosis—determining what is wrong—and treatment. Diagnostic tests employ small amounts of an isotope, limiting the exposure so that little damage results. Treatments make use of larger amounts, most often in the attempt to kill cancer cells.

A frequent need for physicians trying to diagnose an illness is to examine the affected tissue. For tissues beneath the skin, this means either performing surgery or taking images with devices such as an X-ray machine or magnetic resonance imaging (MRI). X-rays are excellent for imaging bones but not as effective for soft tissues. MRIs, which will be discussed shortly, provide high-quality images for many organs, but sometimes a concentration of injected radioactive isotopes works better. The isotopes, usually injected in a vein, can often be made to localize in a specific part of the body. Physicians aim special detectors to capture the radiation emitted by the isotopes, providing a measurement and sometimes an image of the internal tissues. The radioactive isotopes are gradually eliminated by normal metabolic processes.

A common example is the diagnosis and treatment of diseases of the thyroid. The thyroid is a gland in the neck that secretes hormones important for growth and metabolism. The element iodine is essential for thyroid function and is concentrated in the gland. Physicians exploit this by injecting a radioactive isotope such as iodine 131, which collects in the thyroid. (Sometimes other isotopes are used. A common one throughout much of nuclear medicine is an isotope of technetium.) This procedure is one of the oldest in nuclear medicine, having been developed in the 1950s when doctors were first learning how radioactivity works. The test provides information on the shape and activity of the different structures of the thyroid gland. Increasing the radioactivity dose is necessary in the treatment of thyroid cancer.

Radioactive tracers are common in both medicine and research. The introduction section of this chapter described an early experiment by George de Hevesy, and more serious uses of radioactivity to label and trace substances have been developed. A common example is positron-emission tomography (PET), a technique that produces images from the decay of isotopes emitting a particle called a *positron.* PET and the splendid images it generates will be discussed in a later chapter.

MRI is another technique for imaging the inside of the body. The magnetic properties referred to in the name come from hydrogen nuclei, and nuclear physics is vital in MRI. The original name for this technique was nuclear magnetic resonance, but the word *nuclear* disappeared from the name in part because medical personnel were afraid that patients would associate it with radioactivity exposure. MRI works by aligning hydrogen atoms in the body with a strong magnetic *field,* about 40,000 times stronger than Earth's magnetic field. (A field is region of space in which forces act. A magnetic field is a region where magnetic forces exert their effects.) Hydrogen nuclei are protons and behave a little like a tiny bar magnet. With the patient placed in a magnetic field, the hydrogen atoms of the body orient themselves in a certain way. Low-frequency electromagnetic radiation—radio waves—probes the location of the protons. The MRI machine generates a map of these protons—the hydrogen nuclei—in a patient's body.

Why is hydrogen important? Roughly 65 percent of the weight of the human body is water, and every water molecule (H_2O) contains two hydrogen atoms. By studying the distribution of the signals created by the MRI, physicians can often determine the state of health of the body's soft (water-containing) tissues. An MRI image of a tumor in the brain, for example, can be lighter or darker (depending on the tumor and the type of MRI scan) and have a different texture than normal brain tissue.

Nuclear-powered Spaceships

People have derived bombs, electricity, and medical procedures from the nucleus of the atom. Now some people want this tiny powerhouse to carry astronauts to the stars.

Transportation using nuclear power is not new. The world's first nuclear-powered submarine, the USS *Nautilus,* launched in 1955. This vessel made a historic trip in 1958 underneath the Arctic ice cap and was the first of many nuclear-powered submarines in the United States Navy. Some modern surface vessels also have engines based on nuclear reactors.

Motivating the development of nuclear-powered submarines was the desire to stay submerged for extended periods of time, as was necessary for the Arctic trip of the USS *Nautilus.* Submerged vessels can also stay hidden from the enemy. Older submarines had diesel engines, but these combustion engines require air. Underwater travel in these submarines needed bulky batteries that were not powerful and did not not last for a long time, forcing the vessel to resurface repeatedly.

A nuclear submarine has no such limitations. Like the land-based power production facilities, these submarines have a reactor that heats water and makes steam, which drives a turbine. The rotation of the turbine spins the propeller shafts, and the vessel moves through the water. Energy from the reactor also generates electricity for the submarine's equipment and life-support systems, such as the machines that circulate oxygen. Today's nuclear submarines can stay underwater for as long as the food supply and the stamina of the crew hold out.

There was also an attempt in the United States in the 1940s and 1950s to design and build a nuclear-powered airplane. An airplane with nuclear engines could stay aloft for weeks or months, thanks to the huge amount of energy available from a

The aircraft carrier USS *Nimitz* gets its power from nuclear energy. These vessels can operate for as long as 25 years without refueling. *(United States Navy/Photographer's Mate 3rd Class Shannon E. Renfroe)*

relatively small amount of material—no need to keep refilling heavy tanks of gasoline that quickly burn up. At a cost of several billion dollars, engineers produced a design for the plane and a couple of prototype engines. A number of flights tested airborne reactors, although no nuclear-powered flights were ever made. In the end the plans were abandoned, mostly for safety reasons. The shielding required for the reactor was heavy, and fears arose over the potential radiation contamination should the airplane crash.

Fears over nuclear propulsion were not unfounded. Soviet Union nuclear submarine *K-19* suffered a tragic accident with its reactor in 1961. While at sea, 1,500 miles (2,400 km) away from port, a pipe carrying coolant through the reactor sprung a leak. Overheating occurred, threatening a meltdown. Members of the vessel's crew fixed the problem only after entering the reactor itself and exposing themselves to fatal doses of radiation. The heroic submariners managed to prevent a meltdown, at a cost of 22 lives (eight died within days and 14 others over the next few years). Another Soviet Union submarine towed *K-19* to a port and workers decontaminated the vessel. These events formed the basis of the 2002 movie *K-19: The Widowmaker.*

The Soviet navy put *K-19* back into service a few years later and the accident was kept secret for many years. But there have been other accidents, and the hulls of more than a half dozen nuclear submarines rest on the bottom of the world's oceans. Concerned citizens warn of the hazards presented by the reactors still in these sunken vessels, but submarine experts dismiss the danger. The reactors are shielded by thick slabs of material, and the radioactive fuel itself is encased in protective alloys that are highly resistant to marine corrosion. Radiation measurements indicate at most only a slight degree of residual radioactivity in the areas surrounding the wrecks.

In light of the dangers, there is little wonder that many people worry about the use of nuclear power to launch spaceships. All previous launches have used some form of chemical propellant, either solid or liquid, to meet the enormous energy needs of spaceships.

But a major problem with chemical propulsion is the mass of the fuel. Not only do rockets and spaceships have to carry combustible material, but also they must carry the oxygen with which to burn it—the ships must operate in the thin atmosphere of high altitudes or in the vacuum of space. Rocket fuel is heavy; for example, the space shuttle requires several million pounds of fuel at launch. The rockets and spaceships themselves are also heavy, since they are sturdy structures built to survive high speeds and acceleration. After the equipment is added, including astronauts on manned launches, the whole system is quite massive.

Launching rockets with chemical fuel is expensive. Sir Isaac Newton's second law of motion states that the acceleration, a, of a body equals the force, F, divided by its mass, m:

$$a = \frac{F}{m}.$$

When m is large, a is going to be small unless F is also large. According to Newton, to accelerate a heavy object requires a lot of force.

The situation is made worse by the strength of gravity. In order to reach space, a rocket must escape Earth's gravity. The required velocity is called escape velocity and is about 6.9 miles/second (11.2 km/s) near the Earth's surface. This is nearly 25,000 miles per hour! Tremendous thrust is necessary for massive rockets to attain this velocity. The engines of the Saturn V rockets that carried astronauts to the Moon in the late 1960s and early 1970s were capable of millions of horsepower, enough power to create miniature earthquakes when the engines were turned on.

The advantage of nuclear power is clear. The vast quantity of energy locked away in a few ounces of nuclear fuel would be ideal for space launches. Using nuclear fuel, propelling rockets out of Earth's gravitational grip would not be so difficult or expensive.

Several projects have attempted to design a nuclear-powered spacecraft. The goal of the Orion Project of the 1950s was to build a vessel propelled by a series of atomic bombs. In the 1970s the *National Aeronautics and Space Administration* (*NASA*), the United States government agency responsible for space exploration, attempted an equally ambitious plan called Nuclear Engine for Rocket Vehicle Application (NERVA). Neither of these programs was successful.

But NASA has not abandoned the idea. Currently there is a NASA program titled Prometheus Nuclear Systems and Technology that hopes to develop space-based nuclear energy resources. (In Greek mythology, Prometheus gave humanity the gift of fire.) The program focuses on two applications: electrical power generation from radioactive isotopes, and nuclear fission reactors.

Radioactive material has already been used in space probes. But the amount has been relatively small and not powerful enough to propel the craft; the radioactive material simply provided energy to run the probe's instruments. A machine in the probe converts the heat from plutonium 238 decay into electricity, providing enough current to operate all the electrical equipment.

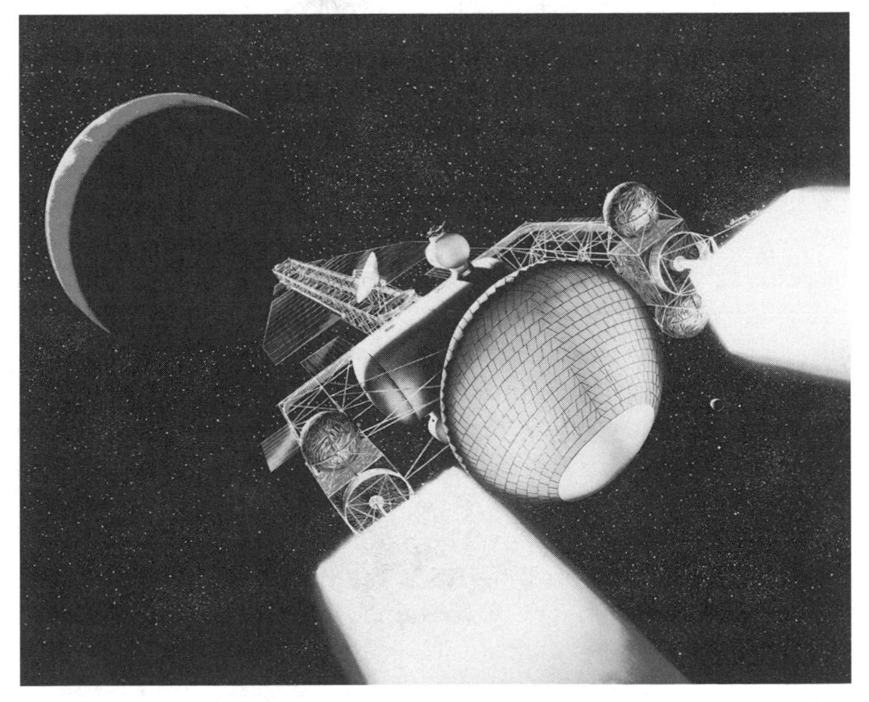

A nuclear-powered spaceship could travel to the outer planets in a few months, instead of the few years it takes using other methods of propulsion. *(NASA/Patrick Rawlings)*

This radioisotope thermoelectric generator is one of three such devices that provide electricity to NASA's *Cassini* probe. These machines produce electricity by converting heat from the radioactive decay of plutonium 238 dioxide. *(NASA-KSC)*

Solar energy is an excellent source of energy in space, and many probes take advantage of it, but for the craft whose missions take them away from the Sun, far into the outer reaches of the solar system, solar energy is not a good option. The Sun, though still relatively bright when viewed from distant orbits like those of Jupiter and Saturn, does not provide much energy to a power-hungry spaceship. Plutonium decays slowly, lasts a long time, and is lightweight. The *Viking* probes of the 1970s, along with more recent probes such as *Galileo* and *Cassini,* have carried plutonium for their electricity needs.

But not everyone is happy. The Saturn-bound *Cassini,* launched in 1997 in a joint effort by NASA, the European Space Agency, and the Italian Space Agency, contained enough plutonium (about 72 pounds [320 N]) to incite protests from concerned citizens. Perhaps some of the concern was based on a confusion over isotopes. Plutonium 239 is a common component of nuclear weapons, but the isotope on *Cassini,* plutonium 238, emits alpha particles and is not terribly dangerous. NASA calculations for the worst-case scenario indicated only a small and acceptable risk.

Even so, the public is not always accepting of assurances by space agencies. There is some historical basis for this skepticism. NASA's space shuttle disasters of 1986, when *Challenger* exploded, and 2003, when *Columbia* crashed, drove home the point that space operations remain dangerous. The use of a small amount of

Cassini, launched on October 15, 1997, to explore Saturn and its moons, ventured too far from the Sun to be powered by solar energy, so it required a nuclear energy source to supply its electricity needs. *(NASA-JPL)*

a well-behaved radioactive isotope such as plutonium 238 to generate electricity on a probe is one thing; a powerful fission reactor on the tail end of a rocket is another, quite different undertaking. Nuclear-powered propulsion for spaceships is technically feasible, but space agencies must convince people that such ships are safe before they begin to appear on the launchpad.

Fusion: Nuclear Power of the Future

The Sun and other stars, like nuclear submarines, get their energy from nuclear reactions. But for astronomical bodies the process does not involve splitting atoms but rather joining them. Stars are powered by fusion.

Fusion releases a proportionally greater quantity of energy than fission—0.3 percent of the mass of hydrogen isotopes is converted into energy when they fuse and become a slightly lighter helium nucleus. Einstein's equation, $E = mc^2$, still applies, so this makes plenty of energy and quite a bang when placed in a bomb.

Because of electrical repulsion, the only way to get protons and positively charged nuclei to come together is to heat them to tremendously high temperatures. Although for most purposes people think of temperature as indications of hot and cold, on an atomic level, temperature is a measure of how fast the atoms and molecules in a body are moving. The atoms and molecules of a hot gas are zooming around with enormous speeds, and the atoms and molecules making up a hot liquid or solid body are jiggling around so much that the body expands. (This is called thermal expansion and is the mechanism by which those old-fashioned liquid thermometers work—warm temperatures cause the fluid to expand and rise.) When hydrogen isotopes in a gas reach a hot enough temperature, they crash into each other with sufficient force to overcome their electrical repulsion. When they get close enough to touch, the strong nuclear force binds them together, and a helium nucleus is born.

The temperature in the center of a star is unimaginably hot. This temperature cannot be measured directly—scientists have no way of inserting a thermometer there—but calculations and theory suggest that the center of the Sun, for example, is about 27,000,000°F (15,000,000°C).

Why is it so hot? The Sun is a huge ball of gas, mostly hydrogen and helium, having a volume about 1.3 million times bigger than Earth and containing more than 330,000 times more mass. The gravitational force of all that mass squeezes the Sun's core with a crushing pressure. This energy heats up the gas at the core, sending the temperature to a level that will support fusion of the hydrogen nuclei. Fusion produces a vast amount of heat and light energy, sending particles and radiation away at high speeds; this energy provides a counteracting force to that of gravity. A star like the Sun is in equilibrium, with gravity trying to squeeze the gas into a smaller ball and the energy of fusion creating a balancing force that tries to swell the gas from within. The Sun has maintained this steady state for about four and a half billion years and will continue to do so for a few billion more.

Later in a star's lifetime it may yield other nuclear reactions, producing elements heavier than helium. In all these reactions a small amount of mass is converted into energy, as given by Einstein's equation. (This means that stars are continuously reducing their mass, but because of the size of the c^2 term the reduction is insignificant.) For many stars the process ends with iron nuclei—these nuclei are the most stable, and further fusion does not emit energy but rather requires it. Elements heavier than iron are made not in ordinary stars but mostly in a still poorly understood event called *supernova,* in which a massive star explodes in a flurry of nuclear activity. There is enough energy in these explosions to produce nuclei of all the heavy elements. Chapter 5 discusses these dramatic events in more detail.

There are several fascinating aspects of star fusion. Because most of Earth and all the organisms that live here contain many elements heavier than hydrogen and helium, the interiors of stars are the crucibles in which the materials of most of the planet and its life were made. All creatures, including humans, consist of elements "cooked" in the stars.

The Sun also provides virtually all the planet's energy, which heats the surface, drives the winds, and gives fuel to plants and, ultimately, to animals. Fusion is a splendid source of energy and if it could be harnessed here on Earth, much of the world's power needs could be met cheaply and safely. The fuel is inexpensive and abundant, as the hydrogen isotope deuterium can be extracted from water for just pennies. Unlike fission, the process that currently drives nuclear reactors, the product of the fusion of hydrogen isotopes is a stable (nonradioactive) nucleus of helium. The fuel required for a fusion reactor's operation would be a safer, smaller amount, decreasing the risk of an accident. One could not ask for a better energy source; it is almost like a dream.

But the dream has a few obstacles. The surface of Earth is nothing like the interior of the Sun, but fusion requires similar conditions or it will not occur. Bombs that employ fusion must recreate the Sun's core by some means, usually by the detonation of a fission nuclear bomb. The fission bomb goes off first, heating the weapon's hydrogen isotopes, deuterium and tritium, to millions of degrees. Then, and only then, does fusion begin. The result is a horrific fireball, as well as scattered radioactive material (much of which is from the initial fission event).

Obtaining constructive rather than destructive energy from fusion requires a controlled process. Herein lies a problem. The heat from fusion is so great that it tends to scatter the fuel. Imagine a wood fire that, once lit, kept blowing itself apart and forcing a cold and unhappy camper to retrieve the sticks the whole night. Gravitational pressure from the Sun's huge mass keeps the hydrogen together, but on the surface of Earth, people must devise other methods.

One possible method, called inertial confinement, makes use of tiny, solid pellets. The pellets contain the hydrogen isotopes to be fused. Bringing about high temperature is the job of laser beams, which are aimed straight at a pellet. The lasers heat the surface quickly, and the surface material pushes against the interior and compresses it. In the process, the center of the pellet reaches the temperatures required for fusion.

Another method confines the fuel by the application of magnetic fields. The high temperatures needed for fusion are hot enough to break all chemical bonds and even remove some electrons from their atomic orbits, and substances at these temperatures are in a state of matter known as a *plasma*—an ionized gas. Because of the electrical charges of the ions, plasmas respond to electromagnetic forces. If the swirling ions in the gas try to escape, their trajectory takes them into magnetic fields that are designed to provide a restoring force. One such magnetic containment device is called a tokamak, which is shaped like a doughnut. (The word *tokamak* comes from the Russian language and refers to the doughnut-shaped chamber and magnetic coils.)

While these methods permit fusion to take place, there is one final requirement for the economic production of power. Before people start to generate electricity from fusion reactors, these reactors must be efficient. At the very least, the machines must produce more energy than they consume. The unfortunate situation at present is that they do not, as the extreme conditions necessary for fusion are expensive. Reproducing the Sun's interior is costly, and the controlled fusion devices of today are not profitable. The problem is not one of making fusion happen; it is one of economics. Physicists have

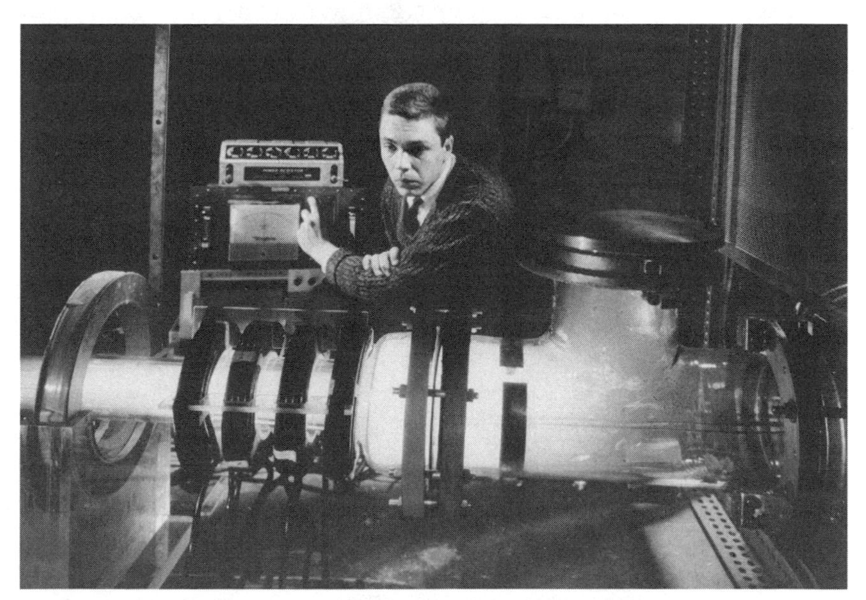

Accelerated by electromagnetic forces, the ions of this plasma move quickly. Plasmas can be used in this way to provide thrust for rockets. *(NASA)*

achieved fusion for a long time, even back in the 1930s, but only on a small scale and with a significant input of energy. Controlled fusion on a large scale, as it is presently understood, needs the heat and pressure of a star's interior to work.

Without those costly requirements, energy from fusion would be easy. Is it possible that some process exists by which fusion occurs under normal conditions on Earth? People are actively seeking such a process by which "cold fusion" can occur, and two scientists, Stanley Pons and Martin Fleischmann, caused a lot of excitement in 1989 when they announced that they had found one. Their simple equipment included electrodes (conductors that carry electric charges) hooked up to a battery and immersed in heavy water (in which the hydrogen nuclei are deuterium) at room temperature. The scientists claimed that their measurements showed a production of energy from a process that they concluded was not from chemical reactions but instead due to fusion.

The news was widely reported in scientific journals as well as in newspapers and magazines. If it were true, energy would become cheap and plentiful. The findings attracted a lot of attention from other scientists, many of whom performed the same experiments but came up with different results (or a different interpretation). The results of Pons and Fleischmann have since become controversial, and many scientists do not believe that these experiments conclusively demonstrated fusion.

Theorists are especially unconvinced. Physicists have spent a lot of time working on the nucleus of the atom and, thanks to the work of Rutherford, Marie and Pierre Curie, Einstein, Fermi, and many others, an experimental and theoretical understanding has been achieved. Cold fusion seems unlikely.

But when one stops to think, a lot of what has been accomplished with the nucleus seems unlikely—cataclysmic bombs, medical devices to peer inside the human body, fission reactors that generate a significant fraction of the world's electricity, all from a tiny package of protons and neutrons held together with the strong force. The ultimate test for any controlled fusion device, whether hot or cold, is the economic production of energy. Perhaps the amazing discoveries arising from the nucleus are not yet finished.

2

Quantum Mechanics

LIKE MANY LOTTERIES, the Pennsylvania state lottery determines the winner of one of its daily games by drawing numbered balls or spheres that are randomly tumbling in a container. The jackpot increases as players buy tickets, and then every day at 6:59 P.M. there is a drawing of the winning numbers. The random selection ensures fairness, since every player has an equal chance to win a share of the money.

But a random selection means that the outcome is unpredictable. Physics is the study of forces and motions, and some people might think that a skilled physicist would be able to calculate the winning numbers by studying the trajectories of the balls. The large number makes the job difficult, yet the container is small—it fits on a little table—and the motion is due to simple causes such as rotation or a jet of air. The trajectory of any ball is complex because it bounces off the wall and other balls, but there is nothing complicated about how or why it moves.

Pierre-Simon de Laplace (1749–1827), a French scientist, believed that physics should be able to predict any motion whatsoever. In the late 17th century, Newton discovered the laws governing the motion of particles, and these laws are universal, applying to all particles in the universe. Laplace and other scientists argued that since these laws govern all motion, a physicist with enough information could predict the evolution of the entire universe. The universe, according to this view, was like a mechanical clock, with the motion of every particle, body, and planet precisely determined and governed by the laws of physics.

This *determinism* made sense to scientists who are accustomed to cause and effect. For example, an object slams into another, causing both to bounce away at predictable speeds and directions. Analyzing the whole group of objects meant that a person could plot the trajectory of every object in advance. But as much as this idea made sense, it proved to be wrong. This chapter describes the stunning development of physics called quantum mechanics that changed the way scientists think. Quantum mechanics banished Laplace's clockwork universe and replaced it with a universe full of *probability,* where there is always an element of chance.

Forces and Motions of Small Particles

Quantum mechanics originated in the late 19th and early 20th centuries, as physicists began to study the atom and its component particles. As described in chapter 1, Thomson discovered the electron in 1896, and Rutherford discovered the atom's nucleus in 1911. Physicists realized that the strong nuclear force held the positively charged protons together in the nucleus, but an understanding of the electrons remained elusive. A simple picture emerged of the lightweight, negatively charged electrons orbiting the heavy, positively charged nucleus, but it was clear from the beginning that this idea did not explain everything.

The biggest problem with orbiting electrons is that they should be continually radiating energy, but they do not—and cannot. When accelerated, electrically charged particles such as electrons emit electromagnetic radiation. Acceleration refers to starting, stopping, or any other change in motion, including turning; orbiting electrons would experience an acceleration continually as they circle the nucleus, so they should be emitting radiation continually. Radiation is energy, and this energy would have to come from somewhere. The only known source would be the energy of the electron's motion or position. An electron emitting electromagnetic radiation would therefore slow down and, as the particle slowed, the electrical attraction between its negative charge and the positive charge of the nucleus would drag it toward the atom's center. As a result, the electron would spiral into the nucleus. The simple picture of the atom was an impossibility—according to physics (as understood by physicists at the time), a stable, long-lasting atom consisting of orbiting electrons could not happen.

Danish physicist Niels Bohr (1885–1962) decided that if the current theories of physics did not work, then he would come up with a new theory. He proposed in 1913 a model of the hydrogen atom, the simplest atom, in which the single electron could only move in certain, prescribed orbits. In Bohr's theory, electrons emit radiation only when they change from one orbit to another. There was little basis for this theory except that it worked. Bohr's ideas proved successful in describing the properties of hydrogen atoms, including the electromagnetic emission *spectrum*—the specific frequencies at which these atoms radiate electromagnetic waves. Bohr received the 1922 Nobel Prize in physics for this work.

Earlier Max Planck (1858–1947), a German physicist, arrived at a similar conclusion while he studied the electromagnetic radiation emitted from heated objects. Planck was trying to understand the spectrum of this radiation, but the experimental measurements did not fit the prevailing theory—Wien's exponential law—at the lower frequencies. Planck could explain the spectrum of radiation in no other way except to assume that the radiation came in packets, or discrete clumps. Until this time, physicists had believed energy was a continuous quantity, similar to the real number line in which there is a number at each and every point. Planck found that energy was instead like the integers, a discrete set of numbers incremented by units. After Bohr postulated his theory of the hydrogen atom, it became clear that the shifts in electron orbit corresponded to the emission of Planck's packets of electromagnetic radiation, and a packet or unit of energy became known as a quantum (Latin for *how much*). The 1918 Nobel Prize in physics went to Planck for his concept of quantized energy.

Around this same time, in 1905, Albert Einstein caused a stir when he proposed that light and other types of electromagnetic radiation consist of particles called *photons*. This idea of photons fit

Wave-Particle Duality

Waves and particles have fundamentally different properties. A particle is bounded—a bundle of mass and energy—and a wave is a periodic motion. A table tennis ball is particlelike, whereas the back-and-forth motion of a guitar string is a wave. The energy of a particle is confined, but the energy of a wave is spread out in the motion that constitutes the wave itself.

Thomas Young (1773–1829), a British physicist and physician, showed that light was a wave when he performed a double-slit experiment. When light passed through two slits in a wall and illuminated a screen, there was a series of bright and dark bands, as shown in the figure on the facing page, instead of two patches of light. If light consisted of particles, the two patches would be expected—the particles would travel through each slit and illuminate two separate parts of the screen. What Young found was that light from each slit spread out and interfered with each other, as waves do. Where one wave's crest aligns with the other's trough, the two waves cancel and a dark spot appears, which explained the dark bands. Light could not be particles because there is no way that particles could combine to produce zero illumination. Scientists considered Young's experiment as proof that electromagnetic radiation is an electromagnetic wave.

But Einstein was willing to consider the alternative. Einstein used the concept of photons to explain the photoelectric effect, in which metals emit electrons when struck by light. The photoelectric effect defied explanation because the metal did not appear to absorb a wave's energy when it emitted electrons. Shining a bright light of low frequency for a long time did not dislodge any electrons, but when light exceeded a certain frequency, the metal began to spew them out. Einstein showed how this could occur if the metal absorbed discrete bundles of light (photons) whose energy depended on frequency—the higher the frequency, the greater the energy. Einstein correctly postulated that the photoelectric effect occurs when a photon with enough energy to liberate an electron hits the metal.

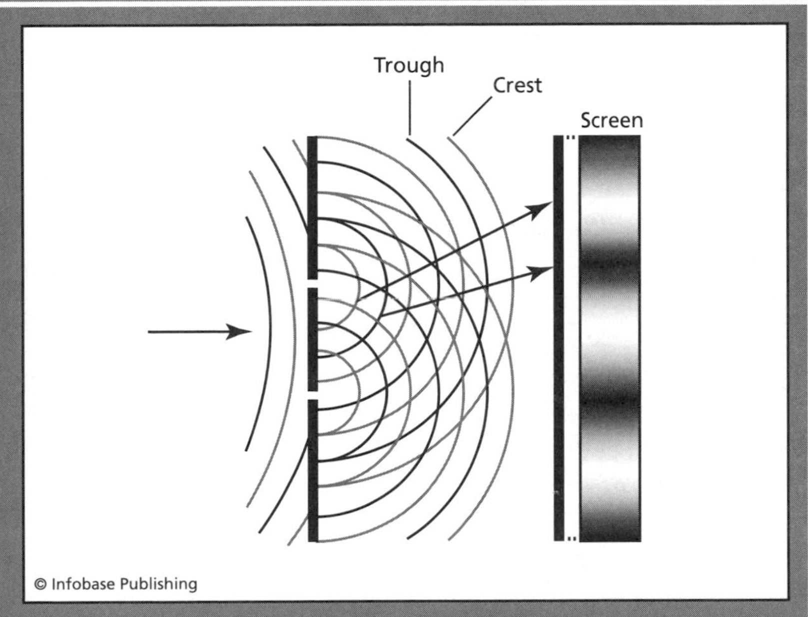

Waves passing through each of the two slits expand and strike the screen. Since they travel different distances, they may arrive at a point on the screen completely out of phase—one wave's crest may align with the other wave's trough. In those cases the two light waves cancel, producing dark bands.

Einstein's photons put physicists in a dilemma. Is light a wave or a particle? Light cannot be both, yet this is exactly what it seems to be, and even photons seem to have a "frequency," a wave property. When performing Young's double-slit experiment using weak sources of light so that only one photon passes through at a time, the screen shows spots of light indicating the existence of photons, but over time the bright and dark patterns emerge as if the photons have interfered with themselves!

Some scientists have entertained the notion of a "wavicle"—a cross between a wave and a particle—but its nature is unknown and is not easy to imagine. Bohr believed that since both wave and particle properties are present, they are both required for a complete description, and he proposed the concept of complementarity. Waves and particles are complementary, each supplying an essential piece of the puzzle.

well with the discrete energy units proposed by Planck; a particle is a discrete, separate entity and can carry a certain amount of energy. If energy is quantized, then the existence of photons makes sense. But as described in the sidebar, this presented a tremendous difficulty. Light was already "known" to be a wave, an entirely different object than a particle!

Bohr's atom, Planck's quantum, and Einstein's photon forced scientists to accept the discrete nature of small bits of matter and energy. This was the beginning of the revolution in physics known as quantum mechanics. But large-scale objects do not seem to be affected. Planets, unlike electrons, do not seem to be limited to certain discrete orbits, since planets can be found in any orbit. Cars, footballs, and Frisbees are also apparently unlimited by the discrete nature of energy and can go at any velocity at all. The reason this seems to be true, though, is that the basic unit of energy is so small that it only becomes noticeable on the scale of atoms and the electromagnetic radiation that they emit. According to theory, the energy of planets and cars also comes in discrete bundles, but because the unit is so small, it is not measurable at this large scale. A drop of water added to the ocean lifts the sea level by a tiny, discrete amount, but not even the most careful islander in the Pacific would be able to detect it.

The wave-particle duality of electromagnetic radiation is a strange and important consequence of quantum mechanics. But quantum mechanics became even stranger after French physicist Louis de Broglie (1892–1987) realized in 1923 that wave-particle duality does not just apply to light. Not only do electromagnetic "waves" have wave and particle properties, but also so do particles.

Electron Microscope

There is a satisfying although mysterious symmetry in the idea that both waves and particles share each other's properties. According to de Broglie's theory, the *wavelength*, λ, of a particle is related to its momentum, p, by the following equation:

$$\lambda = \frac{h}{p}$$

where h is a number called Planck's constant. Momentum usually refers to the product of an object's mass and velocity, but in de Broglie's equation, p is more complicated because it takes into account Einstein's theory of relativity, as described in a later chapter.

No one can "see" the wavelength of a particle, but de Broglie was correct. Particles such as electrons show wavelike properties such as interference, as in Young's double-slit experiment discussed in the sidebar, "Wave-Particle Duality." Objects with a lot of mass, such as footballs, cars, and planets, have vanishingly small wavelengths since p is so large, but tiny particles like electrons have a λ detectable in interference experiments. De Broglie won the Nobel Prize in physics in 1929.

Only a few years later, scientists began to use this strange but true concept to peer further into the microscopic world. Microscopes with optical lenses to magnify the light coming from small objects have long been used by biologists to study cells and bacteria, but light-based microscopes offer a limited resolution. Resolution refers to the ability to distinguish objects; resolving an object under the microscope means that the object can be seen and separated from other objects that may be in the vicinity. Because light spreads out when traveling through the optical system of a microscope, these instruments can have a resolution of no better than 0.000008 inches (200 nm). Physics and the wavelength of light enforce this limit; it is not due to the equipment. No light-based microscope can do better.

But an electron microscope can improve on this resolution because the wave properties of electrons are different. Electrons can have much smaller "wavelengths," and their bending or spreading out is not as severe as that of light, so the smaller wavelength allows observation of the fine details of objects that were obscured when viewed by the relatively large wavelengths of light. Ernst Ruska and Max Knoll at Berlin, Germany, built the first electron microscope in 1931. Due to the smaller wavelengths, the resolution of electron microscopes can be as low as 0.000000004 inches (0.1 nm), several thousand times smaller than that of light microscopes.

There are several types of electron microscope. Transmission electron microscopes produce images by an electron beam passing through the specimen; for this reason, specimens must be extremely thin, typically in slices of 0.0000004 inches (10 nm), and they are often dried or treated with chemicals. Since even air molecules are large enough to deflect the tiny electrons, the chamber of the microscope must be a vacuum or the electron beam would be scattered. (Lenses would not work for the same reason, so electron microscopes use magnetic fields to focus the beam.) The thinness of the sections and the vacuum precludes imaging living specimens, but a magnification of nearly a million times provides a richly detailed view of the sample.

Scanning electron microscopes work by focusing the beam on a small area of the specimen and moving over the surface. The surface scatters the beam, and sensitive detectors create a map, generating a three-dimensional image. Although the specimen must still be enclosed in a vacuum, thin sections are not required and so anything that can survive a short period of time without air is a candidate. (Insects can be imaged, though they must be glued in place so they do not crawl away.)

Electron microscopes require more elaborate equipment than do light microscopes and are more expensive, but they allow scientists to see things that they have never seen before. An important early application of electron microscopes was the study of the smallest living creatures, the viruses. Although there is some debate as to whether a virus is actually alive, the effect of viruses on other living beings is undeniable. A virus consists of a nucleic acid—deoxyribonucleic acid (DNA) or ribonucleic acid (RNA)—wrapped in a coat of protein. Viruses display little activity until they infect a cell of a living creature and take over the biochemical molecules that are used in catalyzing reactions and copying genes. The nucleic acid of viruses contains a small number of genes that are copied, producing many more viruses that go on to infect other cells. Viruses are responsible for devastating human illnesses such as smallpox and, more recently, AIDS and Ebola fever, as well as the many strains that cause influenza and colds. The word *virus* comes from Latin and means *poison*.

The first virus to be found, in the late 19th century, infected tobacco plants, then an important crop for many farmers. Taking samples from diseased plants, Dutch botanist Martinus Beijerinck (1851–1931) carefully applied filters in the attempt to trap the disease-causing agent. If the agent had been a bacterium, the experiment would have succeeded, but the microbe causing the disease was so small that it slipped through the filters.

Most viruses have a diameter of about 0.000004 inches (100 nm), just below the resolution of light microscopes. The study of these tiny infectious agents made little progress until the electron microscope extended the vision of biologists, allowing them to visualize even the smallest viruses. In 1939, for example, the rod-shaped tobacco virus finally showed up in the images of electron microscopes. For most of the 20th century, scientists used electron

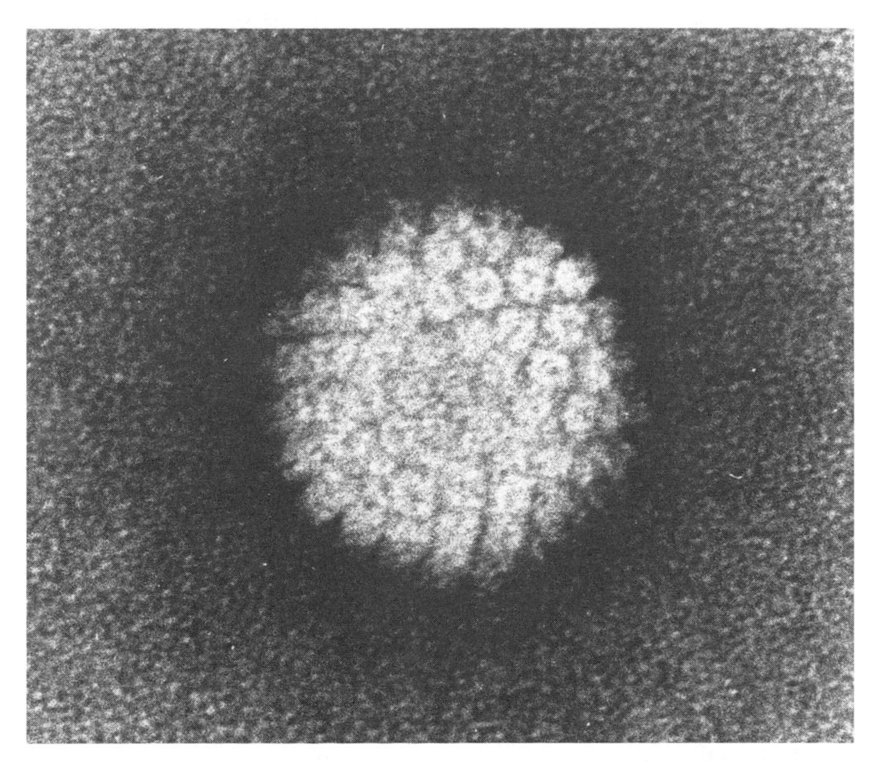

This is an image, taken by an electron microscope, of a human papilloma virus, which causes warts. *(NCI)*

microscopes to identify and study viruses that caused specific diseases such as influenza, polio, and many others; once the viruses were identified, diagnostic tests could be developed to determine who had been infected, and vaccines could be generated to protect against further spread of the disease. Advances in molecular biology and biochemistry, such as devices that can detect and replicate a single copy of viral DNA, now provide many more important tools, but electron microscopes continue to play a role in the fight against viruses and the study of the microstructure of cells.

The Limits of Knowledge

The tiny wavelength of electrons let scientists study and understand viruses, but the notion of a wavelength-bearing particle, even one as small as an electron, continued to mystify them. Physicists eventually came to an understanding of quantum mechanics, but this understanding is more of an admission that parts of the theory will always be difficult or impossible to comprehend. Quantum mechanics imposes a kind of barrier—there will always be a limit on the amount of information people can acquire about any particle or system of particles.

Laplace's deterministic universe had no such limits. To Laplace, all a physicist needed to know was the system's initial conditions—the position of each particle and its initial velocity, if any—and with Newton's laws the future evolution of the system could be determined to any desired degree of precision. There was no uncertainty, there was no motion by any particle anywhere in the universe that could not be calculated.

Quantum mechanics replaced this view. The laws discovered by Newton and other early physicists continued to apply in certain situations—these laws are now sometimes called *classical physics*—but they are not always correct, especially in the atomic domain. A modern astronomer uses the equations of classical physics to compute the trajectory of a planet, but an engineer who designs molecular-sized electronic devices must use quantum mechanics or the device will not work. As for Laplace's viewpoint, the nature of quantum mechanical equations differs from classical equations in a critical way. Newton's laws are deterministic—the behavior they describe is rigidly determined, and the object or system of objects has no alternative pathway. Given a set of forces acting on an object, it can take the one and only trajectory governed by the laws of physics. But in quantum mechanics, the laws are *stochastic,* based on probabilities.

Physicists who began to study atoms and small particles in the early 20th century were not looking for stochastic laws because at the time they believed, as Laplace did, that deterministic laws govern all particles of any size, including planets, grains of sand, and even the miniscule electron. Erwin Schrödinger (1887–1961), a brilliant Austrian physicist, derived an equation (now called Schrödinger's equation) in the 1920s that worked well for particles as small as atoms and electrons, and German physicists Max Born (1882–1970) and Werner Heisenberg (1901–76) developed an alternative method based on mathematical objects called matrices. Both of these techniques used abstract mathematics and were difficult to understand. To the dismay of physicists, the results of these mathematical abstractions did not predict a single trajectory for the motion of a particle or system, but instead specified a probability.

For a given set of conditions in quantum mechanics, a particle may take any of a number of possible trajectories. Instead of determining a single trajectory, the equations assign probabilities to each possible trajectory, and which trajectory the particle actually takes is not known until the event occurs. Quantum mechanics may describe a 60 percent chance that the particle will take path *A,* a 25 percent chance that it will take path *B,* and a 15 percent chance for path *C,* but that is all the information the equations contain. What this means is that if a physicist conducts 100 identical experiments, in approximately 60 of them the particle will take path *A,* about 25 will take *B,* and about 15 will take *C.* Under no circumstances can quantum mechanics reveal which path the particle will take on any particular experiment.

This was much different than the deterministic physics of Newton and Laplace. There was seldom an occasion for probabilities in classical physics, for when all the forces acting on a particle or system were known, the equations specified the exact trajectory.

Probabilities were only necessary when not all the forces were known or when the system was so complex that the exact equations were too difficult to solve.

Since quantum mechanics offered probabilities instead of certainties, physicists wondered if there was some additional information or theories that would provide a complete, more satisfying description of nature. The interference pattern in Young's double-slit experiment provided a good subject for study. The interference pattern was easy to explain in terms of waves, but as mentioned earlier, for weak sources of light the pattern builds up gradually, one speck—one particle—at a time. Logic dictated that each of these particles must have gone through one or the other of the two slits. Quantum mechanics specified the two alternatives in terms of probabilities, whereas classical, deterministic physics would have specified which slit each particle took.

But classical physics failed. The particles, whether photons or electrons or some other particle, are identical and are subject to identical forces as they pass through the slits. Under such conditions they should all take the same path, but they do not. Some particles use one slit, some use the other, and physicists could discover no method of determining which would occur before the event happened. Predictions can do no better than the probabilities of quantum mechanics, and unless someone is looking, the particles appear to go through both slits at the same time, just like a wave!

In 1927, Heisenberg formulated these observations into an idea now called *Heisenberg's uncertainty principle.* Heisenberg argued that there is a limit to knowledge, not because of a lack of information but because of the nature of physics. Consider the process of determining a particle's position and velocity. To make the measurement, a physicist must use some kind of instrument that involves light reflecting from the particle—after all, the particle must be seen to be measured, and being seen requires reflection of light. This kind of disturbance has little effect on a car or a football, but it does have an effect on an electron because absorbing this energy is enough to disrupt the tiny particle's motion. Heisenberg believed that precise measurements can never be made because the very act of measuring causes unpredictable changes that could only be determined by another measurement. But this measurement would in turn cause further changes. Knowledge is limited because measurements are limited.

What Heisenberg proposed is that probabilities are the best that physicists can ever do. The uncertainty principle shows how variables such as position and velocity cannot both be known with certainty. If an electron's position is accurately known, then its velocity must remain indistinct, because a precise measurement of position causes a change in velocity, and vice versa. This means that the complete predictability of the universe as suggested by Laplace would be forever beyond reach. Physicists need to know the initial conditions of a system before exact predictions can be made, but according to Heisenberg this is impossible.

Heisenberg, Born, and Bohr were willing to dismiss determinism, but other gifted physicists of the early 20th century did not give up so easily. What bothered physicists such as Schrödinger was that quantum mechanics seemed to neglect logic. Schrödinger insisted that a photon or electron in Young's double-slit experiment actually does go through one or the other slit and that the failure of quantum mechanics to predict which one meant that it is not a complete theory.

According to the interpretation of quantum mechanics given by Bohr, unless someone is looking, the photon or electron really does go through both slits—the act of measurement determines if these objects behave as waves or as particles. Schrödinger described an experimental condition in which he highlighted logical problems with this interpretation. Schrödinger imagined an experiment in which a cat is locked in a box with a vial of deadly poison. (This is a "thought experiment," not an experiment that would be conducted. Schrödinger's goal was to get people to think.) The vial might break at any time, triggered by a random event such as the trajectory of a particle that in quantum mechanics is indeterminate until measured. No one can see or hear anything in the box, so no one knows if the vial has broken and killed the cat. In Bohr's interpretation of quantum mechanics, in which only the act of looking will determine which path was taken and which state the system is in, the cat must be considered both alive and dead until someone opens the box!

Schrödinger's point was that there could be no such thing as indeterminate states. The cat is either alive or dead whether or not anyone looked. But Schrödinger's argument did not work out as well as he had hoped. Instead of casting doubt on quantum mechanics, what he accomplished was to emphasize the potential basis for a new and astonishing type of computer.

Quantum Computers

Physicists recognized how strange their interpretation of quantum mechanics seemed to be. Everyone accepted the fact that quantum mechanics, unlike classical physics, correctly described the motion and behavior of atomic particles. The only disagreement was whether the stochastic nature of quantum mechanics was a limitation imposed by physics or whether there was another, better theory waiting to be discovered.

Bohr understood that the determinism of classical physics made more sense in the ordinary, everyday world. Effects such as the acceleration of a car always have a cause (in this case, a force provided by the engine), and no one would claim that the car could be a wave or a particle depending on the measurement. Big objects obey Newton's laws of motion, classical physics, and determinism. The idea behind Bohr's interpretation of quantum mechanics is that there is always some uncertainty in every situation, but only when tiny particles such as electrons are involved does the uncertainty become significant. A car has a wavelength (given by the equation on page 46), but it is undetectable by even the most sensitive instruments and irrelevant in the motion of the car. In this view, classical physics is an approximation that works quite well for big objects and poorly or not at all for small ones.

Most physicists today accept either Bohr's interpretation of quantum mechanics or a similar version. The accuracy of quantum mechanics in describing the behavior of small particles is excellent. For example, quantum electrodynamics, a theory based on quantum mechanics that describes the behavior of charged particles, is so precise that some of its predictions accord with experimental results to within one part in a billion. This kind of accuracy would be like measuring the coast-to-coast width of the United States and not being off by more than the diameter of a human hair. Although some scientists do not like the stochastic nature of the universe that is implied by quantum mechanics, they cannot argue with its effectiveness and accuracy.

The notion of indeterminate states would seem to be another idea that is well suited for only small objects or particles—this is what Schrödinger argued when he proposed his cat experiment. But despite Schrödinger's thought experiment, some people wondered if there are applications in the larger, macroscopic world. The wavelength of electrons, for instance, determines the resolution of electron microscopes. Perhaps the existence of indeterminate states could also be of service.

In the 1970s, American physicist Richard Feynman (1918–88), who shared the 1965 Nobel Prize in physics with Sin-Itiro Tomonaga and Julian Schwinger for their work in quantum electrodynamics, envisioned a powerful computer that uses indeterminate states in processing information. Today's computers operate on binary data, consisting of long strings with only two numerals, 1 or 0. The 1 and 0 are called bits and represent logical states that are easy for computers to store and process. In the future, a computer based on quantum mechanics—a quantum computer—may also operate with binary data, but in this case the logical state can be a 1, a 0, or some combination of both.

The indeterminate states described by quantum mechanics give a quantum computer much more flexibility than that of a "classical" machine. The basic unit of a quantum computer has been named *qubit* and is a blend of the binary 1 and 0. The advantage is that such a computer has many more operating states and can process data in parallel—all at the same time, instead of serially, one at a time. Consider an ordinary computer acting on four bits, such as 1010. If each bit must be either a 1 or a 0, the number of distinct four-bit words is 16. But if each bit is not necessarily a 1 or a 0 but some combination thereof, then the number of distinct words can be thousands or even millions. The huge number of states gives quantum computers the same power as that of a computer that can work on millions of bits at the same time. Instead of being limited to a four-bit, 16-state word, a quantum computer has millions of options in the same four-bit space.

Computers functioning with indeterminate states would be so powerful that they could easily break today's security encryption. The basis for many codes is the factorization of numbers with a large number of digits. For instance, 128-bit encryption for secure Web pages uses numbers with 128 bits. Numbers of this size can be as big as 1 followed by 40 zeroes. But for the parallel capacity of a quantum computer, 128 bits or even a lot higher is trivial.

As yet no fully operational quantum computer exists. One of the biggest problems is that the indeterminate states collapse on measurement. When someone is watching, the electron passes through one or the other of the slits, and Schrödinger's cat is either alive or dead. Designers of quantum computers must take into consideration the reading and writing of the data and how this affects information processing. In quantum mechanics, the very act of looking changes the system—this is the basis of Heisenberg's uncertainty principle—so quantum computers will be difficult to operate.

But there has been progress. In 2000, scientists at IBM's Almaden Research Center in San Jose, California, designed a 5-qubit computer consisting of the nuclei of five fluorine atoms. "Programming" consisted of pulsing a small amount of electromagnetic radiation into the system, and the computer was able to perform some simple calculations. Researchers at Almaden continue to work on similar methods, such as a system of ions trapped by electromagnetic forces; the containment isolates the ions so that their indeterminate states can survive for long periods. Each ion corresponds to a single qubit, and lasers read and write information from the system. Current research aims to identify the factors that cause these states to collapse. Such collapse limits the effectiveness of parallel computation.

Beam Me Up: Quantum Teleportation

Erwin Schrödinger was not the only physicist who had difficulty accepting the unusual aspects of quantum mechanics. Albert Einstein, who may have been the greatest physicist of the 20th century, argued that quantum mechanics was in some ways an unsatisfying theory. Along with two colleagues, Boris Podolsky and Nathan Rosen, Einstein proposed a thought experiment that he felt would demonstrate a flaw in the theory. But like Schrödinger's thought experiment, all Einstein accomplished was to highlight the theory's strangeness—and provide a possible basis for teleportation, similar to the "beams" used to transport people and objects across space in the science fiction series *Star Trek*.

The thought experiment involved a concept called entanglement. In quantum mechanics, entanglement refers to a situation in which the states of two or more particles are strongly related, even though the particles may be distant. These situations occur because indeterminate states can involve more than one particle, tying their individual states or trajectories together. For instance, each electron of a pair might be known to pass through different slits in the double-slit experiment described above. Although quantum mechanics cannot tell in advance which electron, E_1 or E_2, goes through which slit, if a measurement indicates E_1 goes through slit A, then E_2 must go through slit B.

Entanglement offers quantum computers a way to correlate the logical states of its components, allowing them to work together. But what bothered Einstein is that entanglement acts across any distance. Even if the particles are separated by miles, or even if they are in different solar systems, a measurement performed on one automatically determines the state of the other. Suppose a chemical or nuclear reaction produces two electrons that fly away from each other, and the nature of the reaction is to produce the electrons with opposite spins—one electron spins up, the other spins down. Quantum mechanics describes the reaction only in stochastic terms. The two electrons and their states are known, but which electron spins up and which spins down cannot be predicted in advance. If a physicist conducts an experiment on one electron and measures an up spin, then the indeterminate state of the system, consisting of the two electrons, collapses into a definite state of spin. This means the other electron's spin state, previously indeterminate, is now determined, even though it might be far away.

This puzzled Einstein because his ideas and theories indicated that no information could possibly travel faster than the speed of light. (Chapter 4 of this book discusses Einstein's theories in more detail.) Yet a scientist could have determined the spin state of one of the electrons by a measurement that might have occurred on another planet or even a distant galaxy.

Despite Einstein's misgivings, experiments similar to the paired electron scenario mentioned above prove that it happens. But the results do not violate Einstein's theories because the phenomenon does not transmit information. Measurement of the first electron determines the state of the second, but there is no way an observer of the second electron could know the measurement took place unless a message, transmitted by conventional means such as radio, arrived from the person who did the measurement. To the second observer, the electron's state is still unknown until a measurement is performed or a message is received from the first observer.

Even though entanglement does not violate any laws, it is yet one more strange consequence of quantum mechanics. If the second electron's spin state is indeterminate until observed, as required by the theory, then somehow it "knew" to adopt the appropriate spin state after someone measured the first electron. This was the essence of Einstein's argument, and it resembles Schrödinger's point with the cat. If the second electron's state is a mixture of up- and down-spin that is not determined until observation, then the first observer's measurement somehow influenced the second electron's state. Einstein had no explanation for this.

No one else has explained entanglement in terms of classical physics, either. Despite this, the process can be useful. Although no information can be transmitted, it can make a copy of something and transport it across space, a procedure called teleportation.

Science fiction writers had already invented teleportation, though they did not have a plausible mechanism for it before quantum mechanics. The idea is to scan an object or person at one location and construct an exact replica at another location. The original is usually destroyed in the process. The person or object does not actually move; the only thing transported is the structure, indistinguishable from the original, embodied in the replica. As illustrated in the figure, a teleporter would resemble a fax machine, except that the process of scanning destroys the original.

Entanglement plays a role because of the long-distance action that puzzled Einstein. An experiment consisting of a reaction or some other event provides entangled particles. By conducting a precise set of measurements, observers at one location determine the state of a subset of the particles. The measurements are made in such a way as to copy the state of one of the particles. The measurement disrupts this particle's state because of Heisenberg's uncertainty principle, but the entangled particles are now in the same state. The original is "destroyed" but a replica exists at another location.

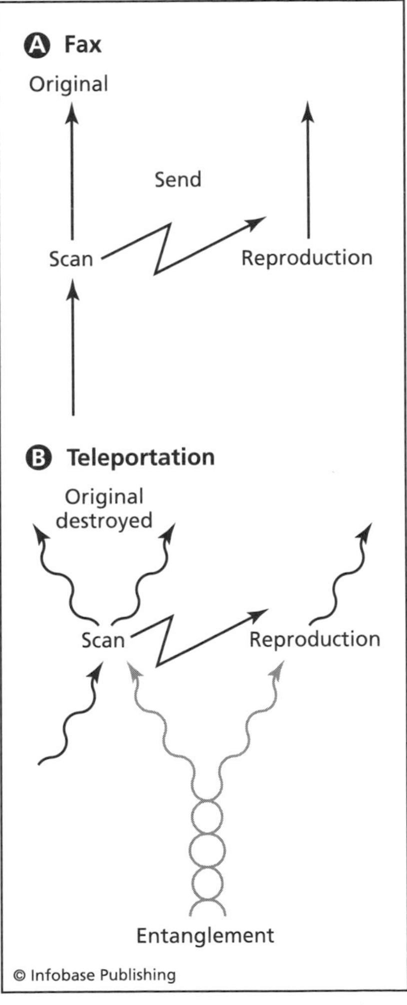

(a) A fax machine scans the original document, transmits a copy over a telephone line, and returns the original. (b) Teleportation "scans" the original, disrupting it because of Heisenberg's uncertainty principle, but entanglement ties its properties to another object, allowing a reproduction to be made over long distances.

Experiments with photons confirm that quantum teleportation can occur. Scientists at the University of Vienna in Austria teleported photons in 2004 across the Danube River. Although photons had been teleported across short distances in earlier experiments, this was the first time the process occurred outside of a laboratory. A fiber-optic cable underneath the river carried bitlike states (qubits) of the photons to the other side.

Another important experiment in 2004 extended the process to atoms. Physicist Daniel F. James of the Los Alamos National Laboratory collaborated with a team of researchers at the University of Innsbruck in Austria, led by Professor Rainer Blatt. Using calcium ions, the scientists teleported quantum states only about a thousandth of an inch, but the experiment is remarkable because atoms are much more complex objects than photons. In order to control the conditions, researchers cooled the atoms to extremely low temperatures and manipulated their states with lasers. The fidelity—the accuracy of the replication—was not always perfect, but the experiment showed that matter as well as light can be teleported.

The present capability of quantum teleportation is a long way from *Star Trek* and is unlikely to be beaming people onto distant planets anytime soon. People are highly complex collections of atoms and a fidelity of less than 100 percent would spell disaster. But the problem is one of engineering. Thanks to quantum mechanics, there is no law of physics that precludes this technology from being developed in the future.

In one bold stroke, physics has made a great deal of science fiction suddenly seem plausible, even likely. Yet perhaps the most important consequence of quantum mechanics is its limitations. Laplace dreamed of a universe that was knowable in every detail; given enough information, every event could be predicted. Quantum mechanics reveals a different universe, where probability plays a role. In Laplace's time, a lottery only seemed to be random—people believed the result was not predictable because of a lack of information. Today, quantum mechanics shows that randomness is not due to a lack of information but is a part of nature. Whether a lottery uses table-tennis balls or electrons, a degree of uncertainty always exists until someone makes a measurement.

3

PARTICLE PHYSICS

STARTING IN 1895 and continuing for several decades, physicists learned many new things. As described in the first two chapters of this book, physicists of this era studied radioactivity, found the atom's components—electrons and the nucleus, composed of protons and neutrons—and discovered that the behavior and properties of small particles are based on quantum mechanics instead of the physics of Sir Isaac Newton. But interesting questions about the nature of these particles remained. People wanted to know more about the forces that electrons, protons, and neutrons exert on one another and whether these particles are composed of even smaller particles.

Finding the answers to these questions was not easy. Microscopes were of little use because the particles are too small to be imaged. Even the electron microscope, discussed in chapter 2, did not help since the beam of electrons it uses for imaging consists of particles having approximately the same size as the particles being studied. To probe electrons and protons, physicists needed an even tinier particle so that they could shoot it at the electron or proton and observe how it bounced off, or made some change in its path. This is the way that scientists use electron beams to study small objects; a similar process would have provided information on the nature and surface of an electron or a proton.

The problem was that no one knew of any such tinier particle. Lacking this, physicists in the middle of the 20th century decided to observe high-speed electrons or protons as they crashed into one another or collided with other particles. The energy of the collision would be large enough to study the forces between the particles and even tear them apart if they consisted of smaller particles. Another way of viewing this process is to consider the "wavelength," *(λ)*, of these particles, which as discussed in chapter 2 becomes smaller as the speed of the particle increases. Particles at high velocities have tiny wavelengths and provide finer details because they have better resolution.

The experiments involving particle collisions succeeded, producing a tremendous amount of information and a bewildering number of particles. Physicists even found a new kind of matter, called *antimatter,* the existence of which had been predicted earlier by British physicist Paul A. M. Dirac (1902–84). These findings were great advances in particle physics, as well as being starting points for ideas and applications that would go on to accomplish a whole lot more.

Particle Accelerators

The requirement for high speed meant that physicists either had to find a natural source of fast-moving particles or accelerate the particles themselves. Some of the earliest equipment used to study electrons were particle accelerators, although physicists did not tend to think of them as such at the time. Thomson discovered the electron in 1896, as mentioned in chapter 1, and many scientists studied this particle with the help of cathode-ray tubes.

A cathode-ray tube consists of a glass enclosure or tube with all or most of the air pumped out. Inside the tube is a metal plate called a cathode and a source of positive charge, such as a positively charged plate called an anode. Heat applied to the cathode causes it to release electrons, and the anode attracts these negatively charged particles because of Coulomb's law, which says that oppositely charged particles attract one another. The electrons accelerate forward and crash into some kind of detector, often coated with a phosphorescent substance that emits light when struck by electrons. Cathode-ray tubes got the "ray" portion of their name because early physicists thought of the electron beam as a ray. These instruments, which often go by the abbreviation CRT, survive today as the picture screen of certain types of television and computer monitors.

But higher speeds were necessary for the particle-collision experiments. Faster particles have more energy, and a greater amount of energy increases the effects of the collision. Scientists found a natural source of particles having sufficient speed, though it was from an unexpected location—space.

Austrian physicist Victor Hess (1883–1964) discovered cosmic rays in 1911. Many cosmic rays can ionize atoms or molecules, meaning that a collision or interaction with an atom or molecule can tear it apart and produce free charges, a phenomenon discussed in chapter 1. Although initially called rays, they turned out to be high-speed particles coming from space (this explains why scientists named them *cosmic,* a term derived from a Greek word meaning *order* or *universe*). Cosmic rays are a steady bombardment of mostly protons traveling at exceptional velocities. Although astronomers and physicists are not sure of the details, they believe that many of these particles arise from exploding stars called supernova events, described in chapter 5.

Having enough energy to tear atoms apart means that the cosmic "rays" are swift enough to use for collision experiments. Earth's atmosphere stops or slows down many of these high-speed particles before they reach the ground, so the best option for physicists was to do their work on mountaintops or to ascend in balloons to an altitude where the air thins out. By observing cosmic rays smashing into or through various substances, physicists discovered strange new particles, such as the muon and the positron (to be described in a later section).

Cosmic rays yielded intriguing results, but physicists became irritated because they had to rely on nature to provide an essential part of the experiment, and nature is not always predictable. What scientists wanted was an instrument to give them whatever particle they wished to use, traveling at the right speed and available on

Cyclotron—R Cancels R

Magnets of the cyclotron curve the pathway of charged particles such as electrons and protons into a circle because magnet fields deflect charges that are moving perpendicular (at right angles) to the fields' lines of force. If the particle's motion is wholly within a field, it follows a circular path because the force of the magnet curves it all the way around. The radius, *R,* of this circle equals the product of the particle's mass, *m,* and velocity, *v,* divided by the strength, *B,* of the magnetic field and the amount of the particle's charge, *q:*

$$R = \frac{mv}{Bq}.$$

The velocity of a particle is the distance traveled divided by the time. The formula for the circumference of a circle with radius, *R,* is $2\pi R$, where π ("pi") is a constant that is approximately equal to 3.14159. For a particle traveling in a circle of radius, *R,* its velocity, *v,* is $2\pi R/t$, where $2\pi R$ is the distance around the circle and *t* is the time required for the particle to finish one full loop. Lawrence realized that if the formula $2\pi R/t$ replaces *v* in the equation for *R* given above, one arrives at the following equation:

$$R = \frac{m}{Bq}\frac{2\pi R}{t}.$$

Multiplying both sides by *t*

$$tR = \frac{m}{Bq}2\pi R$$

and dividing both sides by *R*

$$t = \frac{m}{Bq}2\pi$$

produces the equation for the time, *t,* required for one revolution or "orbit" of the particle:

$$t = \frac{2\pi m}{Bq}.$$

Since "R cancels R," as Lawrence reportedly said in delight when he wrote out the equation, the time does not depend on *R,* the radius of the circle. This is important because as an accelerator boosts a particle's speed, the particle will move in circles with an increasing radius (see the equation for *R* above).

But the time for the particle to make one revolution does not change, so an accelerator can keep giving a periodic boost at a certain instant in the revolution. In this way, an accelerator can give a charged particle a series of accelerations with a small electric voltage, increasing the particle's speed up to the desired level. This method is safer and more convenient than trying to accelerate a particle by using a single gigantic voltage. The figure shows an illustration of the basic components of a cyclotron.

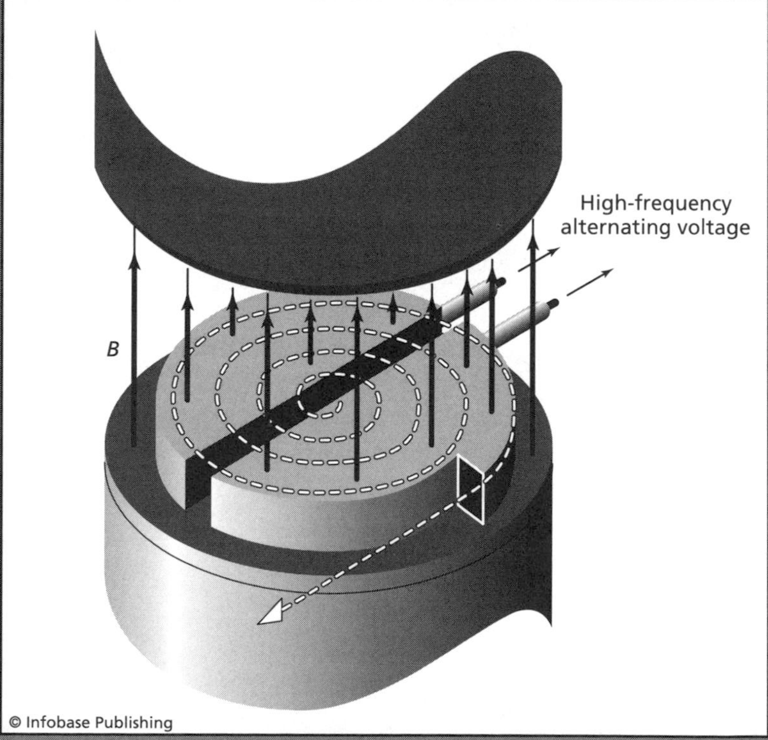

Hollow D-shaped plates called electrodes have a voltage that alternates from positive to negative, the two electrodes always having opposite signs. A magnetic field B *perpendicular to the plane of motion causes the charged particle to travel in circles. The electrode in which the particle is traveling has a voltage of the same sign, and since like repels like, this electrode repels the particle. The other electrode, which has the opposite sign, attracts the particle. The particle accelerates when crossing the gap between electrodes, pushed away by one electrode and pulled in by the other. Then the electrodes switch signs and repeat the process. As the particle gains speed, it spirals outward.*

demand. A particle accelerator would be enormously beneficial because it would permit controlled experiments, repeated as often as necessary. Accelerators similar to CRTs, using greater voltages and distances, could generate acceptable high-velocity particles moving in a straight line, but another method emerged by the middle of the 20th century. These instruments accelerated particles in a circle or ring.

Early circular particle accelerators had their basis in a 1929 invention of American physicist Ernest Lawrence (1901–58). As a professor at the University of California at Berkeley, Lawrence built a particle accelerator known as a cyclotron. The sidebar discusses this important early accelerator.

Hospitals use cyclotrons and similar accelerators to produce particle beams for medical treatments, as described in chapter 1. But because the particles are accelerated within a given area, such as the magnet and the electrode plates as shown in the figure on page 65, they can only be as big as these components.

The biggest particle accelerators of today follow a similar principle of achieving high speeds by a series of steps, although the machine can be either linear or circular. A linear accelerator does not curve the particles into circles but applies a series of accelerations along a straight path. Huge linear accelerators can be found at Stanford Linear Accelerator Center (SLAC) at Stanford University in California and at a few other laboratories, such as Fermi National Accelerator Laboratory (Fermilab) in Illinois. (Fermilab is named for Italian physicist Enrico Fermi, who spent part of his career at the University of Chicago, in Illinois.) SLAC's primary linear accelerator has a length of 2 miles (3.2 km), the longest linear accelerator in the world.

Many of today's circular accelerators consist of a cyclotron, which serves to boost the particles up to a medium-range speed, but they also have a huge ring in which further accelerations occur. Magnets confine the particles within the ring, and rapidly changing electric or magnetic forces increase their speed. These circular accelerators are known as synchrotrons, such as the Tevatron at Fermilab that is 4 miles (6.4 km) in circumference. Synchrotrons have an advantage over linear accelerators since particles can

This linear particle accelerator is located at Fermilab. *(Fermilab Visual Media Services/Reidar Hahn)*

retrace the ring many times, but particles in the linear machine can only go through once.

Even with a huge number of acceleration steps—and with circular accelerators, the number can be as many acceleration steps as the experimenters would like—there are difficulties. Charged particles that are accelerating emit radiation, and the emission decreases their energy. Circular accelerators have a greater problem with radiation because their curved path constitutes a continual acceleration; acceleration is a change in either speed or direction, and the constant change in direction of a curved pathway results in a steady emission.

There are also limits to accelerator performance, imposed by the laws of physics. Particles can never be accelerated past the speed of light in a vacuum, which is 186,200 miles/second (300,000 km/s). Albert Einstein first proposed the existence of this speed limit, and his ideas will be described in the following chapter. An additional problem with circular accelerators is another one of Einstein's discoveries—as a particle increases velocity, it behaves as if it is gaining mass. As a result, the formulas for the cyclotron are not so simple because the time, *t*, depends on mass. This effect requires modifications of the cyclotron concept, achieved in machines known as synchrocyclotrons.

A magnetic core used in a proton accelerator at NASA's Langley Research Center weighs 6 million pounds (26.7 million N). *(NASA)*

The most powerful accelerators today are at Stanford's SLAC, Fermilab, and at the world's largest particle physics center, the European Organization for Nuclear Research, located at the border between France and Switzerland. People also call this center CERN, an acronym derived from an earlier name. CERN houses several linear and circular accelerators and is presently constructing what will be the largest accelerator in the world. Called the Large Hadron Collider (LHC), this accelerator will occupy a circular tunnel 16.5 miles (26.5 km) in circumference, the site of a recently dismantled accelerator. (The terms *hadron* and *collider* are explained below.) Scientists at CERN expect to begin to operate the LHC in 2007.

But it is not just the size of the accelerator that is relevant. The series of accelerations increases the particle's energy of motion, and physicists find it convenient to use the term *electron volt* (eV) to describe this energy. An electron volt is the amount of energy gained by an electron as it moves through one volt. A volt is a standard unit of electrical potential (the ability to move an electric charge) and is comparable to a common flashlight battery, many of which provide 1.5 volts.

Fermilab's Tevatron can give protons an energy equal to nearly a trillion eV, and such high energies create the violent collisions that physicists wish to observe. These particles travel very close to the speed of light. A method of increasing the energy even more is to let two particles traveling at this speed smash into each other in a head-on collision, doubling the energy. Accelerator systems called colliders are designed for this purpose, and Fermilab's collider generates collisions with nearly 2 trillion eV. Plans for LHC include the capacity for collisions of 14 trillion eV. Because of these tremendous energies, people sometimes refer to particle physics as "high-energy physics."

Once a collision occurs, physicists must be able to observe the result. The particles and the "debris" of collisions are too small to be seen by eye or microscope, so detectors employ other means. This usually involves ionization, similar to the process by which Geiger counters detect radioactivity, as mentioned in chapter 1. (Geiger counters were some of the earliest detectors used in particle

A 1999 aerial view of Fermilab shows the Main Injector (a proton synchrotron) in the foreground and Tevatron in the back. *(Fermilab Visual Media Services/Fred Ullrich)*

physics experiments.) Two older types of detector are bubble chambers and cloud chambers. High-speed particles show up in these instruments because of ionization in a liquid for bubble chambers and in vapor for cloud chambers. The liquid in a bubble chamber is close to its boiling point, and when ions form because of the passage of high-speed particles, this causes some of the liquid to evaporate and form easily observable bubbles. In a cloud chamber, the ionization produces clouds similar to the vapor trails of jet airplanes.

Other, newer detectors use electronic devices. Wire chambers are networks of parallel electric wires. The ions created by the ionizing particles flow to the wires, producing a current measured by sensitive meters. The location of the currents in the wire network indicates the track of the particle. Another detector is the vertex detector, shown in the figure below. It is an extremely sensitive instrument consisting of arrays of electronic chips made of silicon. A particle traveling through a chip generates a small amount of electrical charge, and a computer determines the path by examining the location of these charges.

Experiments in particle physics generate a large amount of information. The concept is the same as the scattering experiments with which Rutherford discovered the atom's nucleus, as discussed in chapter 1. Particles collide with each other or with stationary atoms or molecules, and physicists study the identity and tracks of the particles after impact. The energy of the collision is so severe that different particles can form, and the charge, if any, and the mass of these particles are vital pieces of information. Scientists

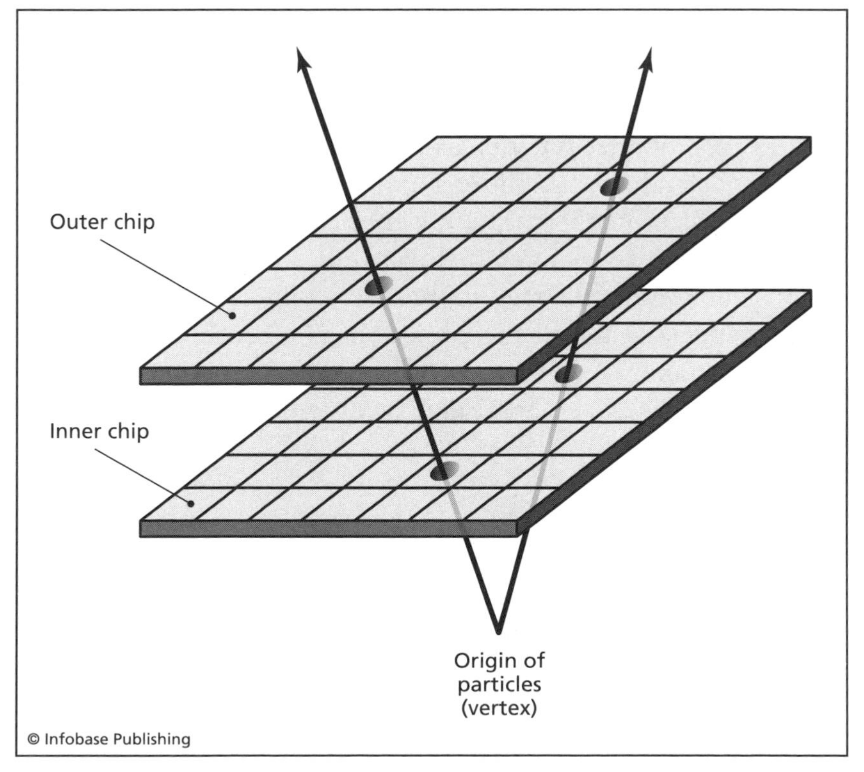

Particles passing through the two layers of silicon chips deposit small electric charges. A computer uses this information to trace the path of the particles. If the particles arise from the same event, the tracks meet at a vertex, or intersection.

measure these properties by various means, such as observing a particle's curved path in the presence of a magnetic field, which as described above depends on the charge and the mass.

All these experiments must be repeated many times because the result of any one experiment is not predictable. Just as two colliding cars may end up in a variety of positions and with different kinds of damage, particle collisions have variable outcomes. Teams of many scientists must study huge quantities of information. Sharing this information among large numbers of scientists was such a burden that in 1989 Tim Berners-Lee, a scientist at CERN, created a method to organize information and make it easily accessible across computer networks. This system was so successful that many people decided to connect, and it became the World Wide Web. The first Web site in the United States belonged to SLAC.

Besides being the birthplace of the Web, particle accelerators are the generators of beams that are used for medical procedures such as radiation treatments, as described in chapter 1. But the largest accelerators cost billions of dollars and require hundreds or thousands of workers. There were plans in the United States to build a huge accelerator in Texas, but after the initial phases of construction ran through more than a billion dollars, the government decided to kill the project in 1993 rather than spend any more money. Yet the scientific accomplishments of particle accelerators have been great and include a richly detailed view of matter.

What All Matter Is Made Of

Previous chapters described how physicists discovered that atoms consist of electrons "orbiting" around a nucleus made of protons and neutrons. Cosmic-ray experiments produced a variety of previously unknown particles, and beginning in the middle of the 20th century, powerful accelerators yielded an even greater abundance of particles. These particles could be distinguished by differences in mass, the sign of their charge, and several other properties. Most of these new particles had short lifetimes, combining with other particles or decaying into a more stable (longer-lasting) particle a fraction of a second after the collision that created them. (An unstable particle is like an unstable nucleus, which decays into more stable nuclei.) Physicists compiled a catalog of several hundred of these particles, and at first no one knew what purpose most of them served. Some people began to refer to this phenomenon as the "particle zoo."

The initial approach to this situation was an attempt to classify the particles, just as biologists in the 17th and 18th centuries began to classify animals. Physicists assigned particles to different categories based on mass—the lightest particles, such as electrons, positrons, and muons, belonged to a category called lepton, and the heaviest particles, such as protons and neutrons, were baryons. (The names of the categories come from Greek terms, *leptos* meaning *small* or *light*, and *baryon* meaning *heavy*.) Particles with masses between these two categories became known as mesons. The term *hadron* refers to both mesons and baryons.

Another important property is the sign of a particle's charge, if it has one. American scientist and statesman Benjamin Franklin realized that charges come in two varieties, which he named positive and negative. Electrons are negative and protons are positive, and it is a simple matter to determine the sign of a particle by observing its interaction with magnets or electric charges. Particles with no electrical activity are neutral, such as neutrons. (These particles are also impossible to accelerate in the machines discussed earlier, although it is possible to obtain beams of high-speed neutrons by various other means, such as combining them temporarily with charged particles.)

Spin is another property of particles. This property got its name from early physicists who thought of it as a rotation, although this proved not to be true. A particle's spin can only be described by quantum mechanics, the subject of chapter 2. Spin is quantized—it exists only in certain amounts and certain directions.

Many people began to think of these particles as elementary or fundamental, the most basic units of matter. Although this notion was appropriate, there was a suspicion that some of these particles were composed of even smaller particles. In the early 1960s, two physicists, Murray Gell-Mann (1929–) and George Zweig (1937–), came up with the idea that hadrons, the heavier particles, were made of various combinations of a small set of tiny particles. This idea was tremendously useful because it reduced the bewildering number of particles down to a more manageable set of fundamental units, enabling physicists to make some sense of their particle "zoo." A model developed out of Gell-Mann and Zweig's idea and other theories involving the forces and interactions between particles. The last section of this chapter describes this model, called the *Standard Model.*

Before the Standard Model appeared, an interesting particle called the positron occupied a lot of people's time and thoughts. American physicist Carl Anderson (1905–91) discovered this particle in 1932 while studying tracks in a cloud chamber. During a cosmic-ray experiment, Anderson observed a particle with the same mass as an electron but having a positive charge rather than a negative charge. This particle, the positron (**posi**tive elec**tron**), proved to be the first encounter with substances that came to be called antimatter. Anderson and Victor Hess, founder of cosmic rays, shared the 1936 Nobel Prize in physics for this discovery.

Antimatter

Dirac proposed the existence of the positron because this particle appeared in the solutions to complex equations he was studying. These equations were associated with the behavior of electrons, but some of the solutions called for a particle of the same mass as the electron but with a positive charge. Anderson found the positron only a few years after Dirac formulated his theory.

The theory turned out to be more general, involving not only the electron but also all particles. Every particle has a "twin" with identical mass, spin, amount of electric charge, and other properties, except that the sign of many of these properties is reversed. *Anti-* is a prefix denoting opposed to or opposite, so these twins became known as *antiparticles.* A positron is the antiparticle of an electron, and an antiproton is the antiparticle of the proton. (Antineutrons are electrically neutral, as are neutrons, but are the opposite of neutrons in other ways, such as in magnetic properties.) Positrons, antiprotons, antineutrons, and the other antiparticles all fall in the category of antimatter.

Earth is made of matter, not antimatter. People quickly discovered the reason that antimatter does not exist in the world outside of particle physics experiments: when a particle and antiparticle meet, they disappear in a burst of energy. As described in the sidebar on pages 76-77, this process is known as *annihilation* and results in the destruction of both particles. Taking their place are highly energetic photons of electromagnetic radiation.

Another way of looking at matter-antimatter annihilation is that a particle and its antiparticle transform into another type of particle, photons, having tremendous energy. Photons with enough energy can also transform into a particle-antiparticle pair, a process known as matter-antimatter production. Photons are extremely adaptable and have another curious feature—physicists believe that the antiparticle of a photon is a photon. In other words, photons are their own antiparticles.

The production of positron-electron pairs requires an energetic photon. Since the mass of a proton and its antiparticle, the antiproton, is about 1,836 times greater than the electron and positron, their production needs 1,836 times more energy (approximately 1.876 billion eV). Production of proton-antiproton pairs requires a tremendously energetic gamma-ray photon and are not as commonly generated as positron-electron pairs. But the collisions in powerful accelerators such as Fermilab's Tevatron generate more than a trillion eV, enough energy to create even heavy particle-antiparticle pairs.

Antimatter is difficult to make and even more difficult to store for any length of time. Contact with matter results in the annihilation of antimatter (along with an equal mass of matter), so antiparticles must be kept away from their respective particles. Since everything on Earth is made of matter, including boxes and containers, this presents a difficulty. Particle physicists maintain antimatter by confining it with *electromagnetism,* in the same way that accelerators confine or focus beams of charged particles. Since magnetic fields deflect charges, an appropriate configuration of

Matter-antimatter Annihilation

The annihilation of matter and antimatter seems at first to contradict the law of conservation of mass, which states that the amount of mass does not change in chemical reactions. But as discussed in chapter 1, this law requires an amendment because of Einstein's equation relating energy, *E*, mass, *m*, and the speed of light, *c*:

$$E = mc^2.$$

The first chapter of this book described how nuclear reactions such as fission and fusion convert a small amount of mass into an enormous quantity of energy, thanks to Einstein's equation and the huge magnitude of *c*, the speed of light. When multiplied by the square of *c*, 186,200 miles/second (300,000 km/s), even a small mass, *m*, results in a large energy, *E*. Reactions and transformations often result in a change in mass, but the combined total of mass and energy remains the same before and after.

A similar conversion of mass into energy occurs in matter-antimatter annihilation, except on an even grander scale. Nuclear reactions usually convert less than 1 percent of the matter involved, but the annihilation of a particle and its antiparticle often transforms the entire mass of both into energy. In terms of electron volts, the mass of the electron plus the mass of the positron multiplied by the square of the speed of light yields 1.02 million eV. If the particles are moving—that is, have kinetic energy—then this energy will also go into making

magnets can deflect antiparticles before they reach the wall of their container and suffer annihilation.

Antiparticles can get together to form atoms, or rather "anti-atoms," in the same way as normal matter. CERN researchers in 1995 created the first antihydrogen atom, consisting of a positron "orbiting" an antiproton. But antihydrogen atoms, at least those produced in the lab, do not seem to be stable and do not have long lifetimes.

People have often wondered if there is any antimatter, and even anti-atoms, floating around in the universe. The Alpha Magnetic Spectrometer (AMS) is a space-based instrument designed to look the photons. The result is photons that have enough energy to be gamma rays, the most energetic radiation in the electromagnetic spectrum.

The process can also go in the other direction, in which case energy of 1.02 million eV transforms into an electron-positron pair, as shown in the figure. This is an example of particle-antiparticle production, the conversion of energy into matter (plus its antimatter twin). Such an event sometimes occurs when a gamma-ray photon collides with an atom's nucleus, and some or all of its energy goes into the creation of a particle and its antiparticle.

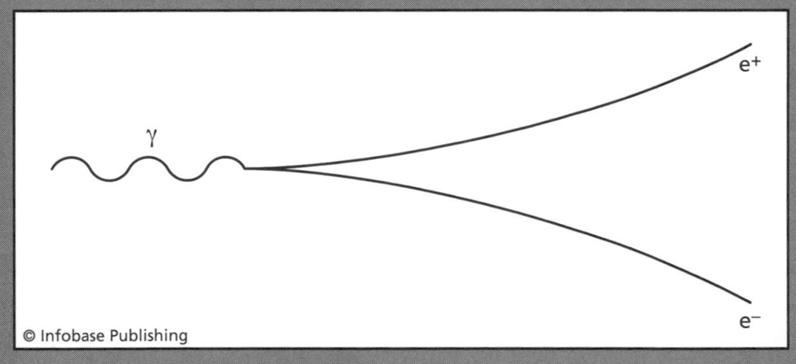

Sometimes an event such as a high-energy gamma-ray photon (γ) striking a large nucleus produces an electron (e^-)-positron (e^+) pair. A magnetic field (not shown) perpendicular to the plane of the figure causes the two charged particles to deflect, but in opposite directions since they have opposite signs.

for antimatter in cosmic rays. An early version, AMS-01, flew on the space shuttle *Discovery* in 1998, and it failed to detect any antimatter. Researchers at a large number of universities and institutes collaborated on the construction of a larger, improved device, AMS-02, and scheduled it for launch in 2005. The goal calls for a space shuttle to carry the instrument to the *International Space Station,* a manned satellite orbiting Earth, where AMS-02 will operate for several years. Problems with the space shuttle fleet caused the 2005 date to slip, but researchers hope to begin the experiments within the next few years.

Few people expect AMS-02 to find much, if any, antimatter. There could exist antistars, antiplanets, and even antigalaxies, but the universe seems to be made predominantly of matter. Any antimatter in the universe may have long since disappeared, becoming annihilated when it came into contact with matter. Although no one knows for certain, some people believe that at the time of the universe's creation there was a slight excess of matter over antimatter. Most matter and all antimatter annihilated each other, leaving a small amount of matter to go on to fill the universe. Why there was an excess of matter and whether this theory is correct are open questions. (Chapter 5 of this volume discusses the universe, including its creation.)

Even though there may not be any anti-atoms in the universe, antiparticles do exist in physics laboratories as well as in a number of important technologies. One application of antiparticles is positron-emission tomography (PET).

Imaging the Body with Antimatter

Physicians and researchers use PET to make maps or images of the inside of the body without having to perform surgery. The annihilation of positrons and electrons is a critical component of PET operation. In the first step, the patient receives an injection of a substance containing radioactive atoms. This substance is often a modified sugar, which the body breaks down, or metabolizes, to provide energy for its cells and tissues. When injected into the bloodstream, the radioactive sugar molecules accumulate in areas of the body that are active and require a lot of metabolic activity. While there, some of the radioactive atoms decay.

As described in chapter 1, the nuclei of radioactive atoms decay and emit a variety of particles, the type of which depends on the specific nucleus. PET procedures employ radioactive nuclei that emit positrons. Examples are carbon 11, oxygen 15, and fluoride 18. The molecule injected for PET imaging incorporates one of these radioactive atoms; for example, the molecule will have an oxygen 15 atom in its structure where it would normally have a stable oxygen 16 atom. This is similar to radioactive labeling

The Alpha Magnetic Spectrometer 2 (AMS-02), being inspected here at NASA's Kennedy Space Center, will search for signs of antimatter in space. *(NASA-KSC)*

studies, also described in the first chapter. The PET radioactive elements have half-lives that are not too short and not too long, so most of the nuclei will decay soon after reaching sites of active metabolism, but not before.

The emitted positrons do not travel far before encountering an electron, and when this happens they undergo annihilation. This event tends to produce two identical gamma-ray photons traveling in opposite directions, as illustrated in the figure on page 80.

Annihilation sends the two photons out in opposite directions because the momentum of such a pair is zero, and there is a law of physics that requires the momentum before and after a transformation to be the same. Photons have no mass, and momentum is usually defined as the product of mass and velocity, but photons do have momentum because they have energy. (By Einstein's equation, $E = mc^2$, photons have an equivalent of mass.) Each of the pair of identical photons created by the annihilation event has the same momentum, but their sum is zero because they are moving in opposite directions. This must be true because in most PET events, neither positron nor electron has much speed and therefore their total momentum is little or none.

The shape of a PET machine resembles a doughnut, with photon detectors situated in a ring around the patient's body. Detectors search for photons of an energy expected for a positron-electron annihilation and traveling in opposite directions. The machine measures the time each pair of photons arrived at the detector and then uses this information to calculate their point of origin. By locating the origin of the annihilation event, PET determines roughly where in the body the radioactive molecule is present. (This calculation is not precisely correct since the positron traveled an unknown distance before meeting an electron, but this distance would rarely be very far.) Regions of the body with greater

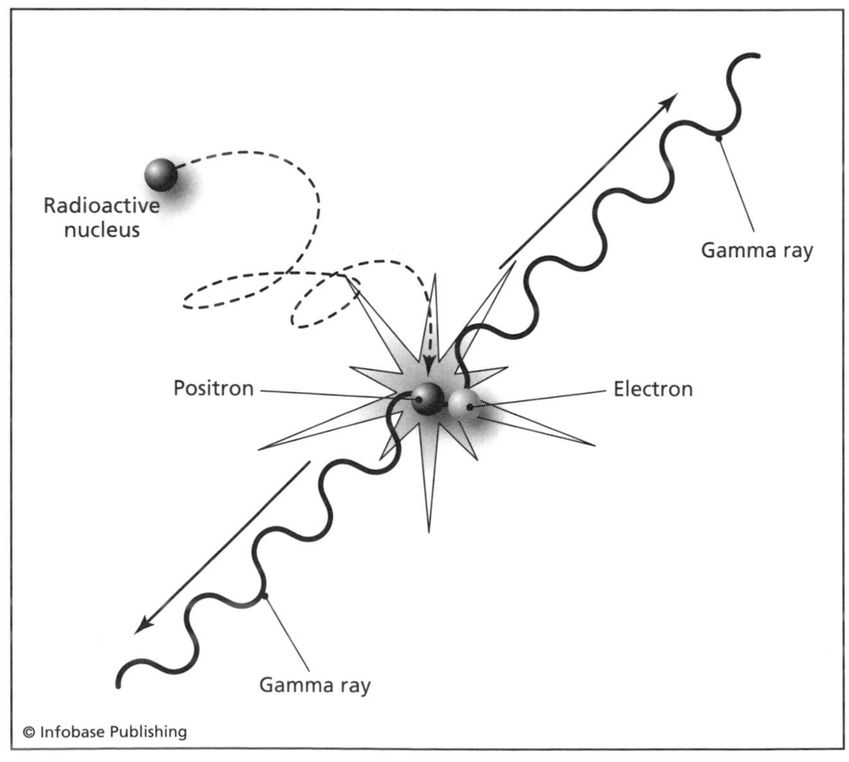

The radioactive nucleus decays, emitting a positron. The positron roams for a short distance before encountering an electron, and the two annihilate each other, producing two gamma-ray photons.

metabolic activity will emit more gamma rays, and PET generates maps showing the level of activity for areas under study.

Physicians often use PET to find tumors (abnormal growths). The cells of these tumors grow and divide quickly and in so doing use a lot of energy. Metabolic activity fuels this rapid growth, and PET excels at locating such high levels of activity. PET is a splendid tool permitting physicians to spot abnormal tissues without having to cut open the body, saving the patient a painful procedure and the risk of infection.

Researchers use PET to study the metabolic activity of organs and tissues. A favorite target of PET research is the brain, an organ with a high metabolic rate. Brain cells become more active as they process sensory information, make decisions, and perform other tasks that lead to the still-mysterious processes underlying human thought and consciousness. Scientists who use PET to image a person's brain can get a series of snapshots of what part of the brain is most active. These images have given scientists a better though still incomplete understanding of how the brain works. For instance, PET images of a person who is viewing pictures shows a high level of activity in a specific region of an important brain structure known as the cerebral cortex. The experiment indicates the importance of this region in vision.

Propelling a Spaceship with Antimatter

Medicine and brain research are not the only applications of antimatter. Particle-antiparticle annihilation is the most efficient method known of obtaining energy—up to 100 percent of the mass becomes energy. This beats nuclear energy by a wide margin and is billions of times more efficient than the burning of fuels such as gasoline. People have proposed spaceships powered by engines that derive their energy from matter-antimatter annihilation. Although such engines are common in fictional ships such as the *Enterprise* in *Star Trek* movies and television shows, they do not yet exist. But the proposals are not at all far-fetched.

NASA's Institute for Advanced Concepts provides funds for research into new rocket-propulsion methods such as antimatter.

The most common antiparticles considered for these engines are positrons and antiprotons. Positron-electron annihilation generates lower-energy gamma rays than proton-antiproton annihilation, but this might be preferred since high-energy gamma rays are exceptionally dangerous and hard to use. Yet photons of any energy level can be difficult to harness. Proton-antiproton annihilation tends to be a more complicated process and produces particles in addition to gamma-ray photons, but these particles may be more useful in developing engine thrust.

In either case, antimatter can be confined with electrical or magnetic methods, and annihilation releases a huge amount of energy. The biggest obstacle for antimatter propulsion is obtaining the antimatter. There is no known source except for the tiny quantity released during radioactive decay or created during the collisions in the center of powerful particle accelerators such as Fermilab's Tevatron. The cost of producing this antimatter makes it one of the most expensive substances on Earth—about $1,100 trillion an ounce ($40 trillion/g). Although only a small amount of antimatter would be required for a quick trip to Mars, it would still cost billions of dollars.

Newer, bigger particle accelerators such as CERN's LHC can increase the rate of the world's antimatter production, but even with these machines, antimatter will be scarce. Starships powered by matter-antimatter annihilation are still in the realm of science fiction, but the concept is valid. If a rich source of antimatter can ever be found, humans may well find themselves riding to the stars.

The Standard Model of Fundamental Particles and Their Interactions

Antimatter and its applications are an important aspect of particle physics, but scientists want a more general concept or theory describing particles, antiparticles, and the forces between them. The most widely accepted theory today is known as the Standard Model. Although referred to as a "model," a term that may suggest a hesitant acceptance, experimental evidence has confirmed the

Standard Model many times since physicists developed the idea in the 1970s.

Critical components of the Standard Model are particles called *quarks.* (Gell-Mann, one of the first scientists to propose the existence of these particles, named them after a word found in James Joyce's 1939 novel *Finnegans Wake.*) According to the theory, quarks make up all hadrons—baryons are composed of three quarks bound tightly together, and mesons are composed of two quarks. The mathematical formulas describing quark behavior are complicated, but the idea, as eventually developed by Gell-Mann, Zweig, and other physicists, simplified the complexity of the "particle zoo."

There are six quarks (and their antiquarks), and various combinations of these particles make up the hadron family, which includes protons and neutrons. Quarks simplified particle physics in the same manner that Russian chemist Dmitri Ivanovich Mendeleyev simplified chemistry in 1868, when he discovered a way to order the elements such that they formed a table, now known as the periodic table. To chemists, elements are the fundamental units of which compounds are made; to particle physicists, quarks are the fundamental units of which hadrons are made.

Some people were skeptical about quarks at first, particularly due to the unusual magnitude of their electrical charge. Electrical instruments are extremely sensitive, and physicists can measure the charge of particles and substances with a great deal of precision. It had been known ever since the early 20th century that electrons and protons have the same magnitude of charge (though it is opposite in sign), and this is the smallest magnitude ever found. Theory required quarks to have a fraction of this charge—some quarks have one-third of the electron's (and proton's) magnitude of charge, and some quarks have two-thirds. But no particles had ever been seen with this charge when Gell-Mann and Zweig first put forward the quark concept, and this continues to be true today. No free, isolated quarks have ever been found. They only exist inside baryons and mesons.

Despite the absence of direct experimental evidence for quarks, physicists are convinced these particles are real. The theory predicted the existence of new hadrons that were subsequently discovered, strengthening belief in the correctness of the theory. As the Standard Model emerged, its formulas successfully predicted properties of the particles and outcomes of collision experiments performed with the large accelerators. Most people have confidence in the Standard Model and the existence of quarks.

Although no accelerator experiment has ever produced a quark in isolation (outside of a hadron), there is a good reason why. Holding quarks together inside hadrons is a powerful force known as the strong force (also known as the strong nuclear force), discussed earlier in chapter 1. To break a quark loose from its confinement would require tremendous energy, well in excess of even the trillions of eV that modern particle accelerators can produce.

The names of the six quarks are somewhat whimsical: up, down, strange, charm, top, and bottom. But quarks are not the only fundamental particles. Leptons, a class of particle mentioned earlier, are not made of quarks but rather exist on their own. There are six leptons: electron, muon, tau, and three types of a particle called the neutrino. The muon and tau have similar properties to those of the electron but are more massive. Only the electron is a common component of matter. Many processes such as nuclear fusion in the Sun produce large numbers of neutrinos, but these particles rarely interact with other matter and are therefore difficult to detect. The mass of neutrinos and whether they have any mass at all has been the subject of debate over the years, but many people now believe that these elusive particles do have a tiny mass, although no one is sure exactly what amount.

Describing the fundamental particles is only part of the Standard Model. Another important aspect of the theory is how particles interact. These interactions give rise to forces, of which there are four types—electromagnetic, weak, strong, and gravitational. Chapter 1 discussed the strong force and the weak force (also known as the weak nuclear force), since these forces are involved in nuclear reactions. Electromagnetic force includes the forces exerted by magnets and electrical charges on one another. Gravitation is an attractive force between pieces of matter.

Of the four forces, gravitation is by far the weakest, much smaller in magnitude than even the "weak" force. The other three forces are so much more powerful that gravitation's effects are insignificant and ignorable on the tiny scale of particles. This is true only on these tiny scales. The weak and strong force only act over short distances, since their strength decreases rapidly over distance, but gravitation, though not as powerful, decreases less rapidly with distance. A pair of quarks in a meson are practically side by side, and at this short range the strong force is vastly greater than the gravitational attraction between these particles. But the magnitude of the strong force drops to zero over larger distances, at which point gravitation becomes the predominant force between uncharged pieces of matter. (Electromagnetism decreases with distance at the same rate as gravitation, and being a more powerful force, it is more important than gravitation for charged objects.) Since gravitation can be neglected on the small scale of particles, the present version of the Standard Model deals only with the strong, weak, and electromagnetic forces.

People often think of a force as a push or a pull, and this simple concept is a valid way of thinking about a force. But particle physics describe forces in terms of an exchange of particles. This is not an easy idea to grasp. One of the first people to put the idea forward was Japanese physicist Hideki Yukawa (1907–81) in 1935. As an analogy to Yukawa's idea, consider the chemical bonds formed by chemical elements that share, or exchange, electrons, such as the powerful bonds that hold two hydrogen atoms to an oxygen atom and form water, H_2O. Yukawa thought that the exchange of a then-unknown particle might underlie the "bonding" of the strong force. This idea is the foundation of the modern understanding of forces as described in the Standard Model, and Yukawa won the 1949 Nobel Prize in physics for his efforts.

According to the Standard Model, a special group of particles carries or transmits all the forces—the interactions—between particles. Each force has its own force-carrying particle. Gluons carry the strong force (gluons are the glue that holds together quarks), bosons carry the weak force, and photons carry the electromagnetic force. Forces occur between particles because they exchange these force carriers, somewhat like a game of pitch and catch. How this translates into a "push" or "pull" is not quite clear in terms of simple, everyday concepts, but the mathematical calculations and predictions of the theory match well with the results of accelerator experiments.

Although there is much evidence for the existence of these force-carrying particles, they are difficult or impossible to detect. Ordinary photons are detectable, but the photons underlying the transmission of a force are called virtual photons. A virtual particle pops out of nowhere, a phenomenon that the conservation of mass and energy would not seem to permit because there would be a creation of mass or energy. Yet the effects of virtual particles are real, and they owe their shadowy existences to an uncertainty, known as Heisenberg's uncertainty principle, described in chapter 2.

Heisenberg's uncertainty principle says that certain pairs of measurements can never be precise at the same moment. The position and momentum of a particle is one such pair, and the more accurately the position of a particle is known, the less accurately its momentum can be known. This trade-off means that a person can know exactly where a particle is located, but if so, the momentum of the particle must be completely unknown. Or, in the opposite case, a person can know the precise momentum of a particle, but if so, the particle's position must be completely unknown.

Another pair of measurements that experiences uncertainty is energy and time. This means that the lifetime of a particle and its energy cannot both be completely certain. As a result, extremely short-lived particles such as virtual particles can pop into existence if they disappear almost at once. No one can observe the mass or energy, so there is no violation of the law of conservation of mass and energy.

Force-carrying gluons and bosons are also virtual particles. The whole concept of virtual particles is strange, for people are accustomed to physics on the scale of everyday life, in yards or meters. This is the same situation people face when considering quantum mechanics. Physics at the scale of particles introduces new ideas and objects that are not easy to grasp, at least not with minds that developed in a world where rocks impact and chip one another, rather than being particles transmitting forces by exchanging virtual particles. Yet the mathematics of the theory is correct, and evidence accumulated with particle-accelerator experiments supports it.

But the Standard Model is not the end of the story. Even though gravitation can be ignored at the particle level, physics is not complete without an understanding of this force. As of now, physicists do not know how to fit gravitation into the accepted theories governing particle physics, although most people believe that there is a particle, called the graviton, to carry this force, similar to the other force-carrying particles. Yet studies of the force of gravitation have tended to go off in a direction different than that suggested by the Standard Model. The following chapter pursues this topic.

As for particle physics, building and maintaining huge accelerators with which to do bigger and better experiments is expensive. Particle physicists compete with other scientists for money given by government and private agencies, and despite applications such as particle beams for cancer-fighting treatments and PET imaging, particle physicists sometimes find themselves on the short end of the budget. Yet particle physics has revealed the fundamental nature of matter and its forces, and the high energies associated with accelerator experiments are an essential tool in the study of the universe. Physics at a small scale can and does produce big results.

4

RELATIVITY

SCOTTISH PHYSICIST AND mathematician James Clerk Maxwell (1831–79) found a way in the 1860s to unite the physics of electricity and magnetism into a single set of equations. These equations showed how electricity and magnetism are deeply interconnected, and the formulas indicated the existence of electromagnetic waves traveling at a fixed speed. When Maxwell realized that this speed was the same as the speed of light, he understood that light is an electromagnetic wave. He also predicted the existence of other electromagnetic waves, and a few years later German physicist Heinrich Hertz (1857–94) discovered radio waves, which are also electromagnetic.

A puzzling feature of these electromagnetic waves is their fixed speed. Light and other electromagnetic waves travel at 186,200 miles/second (300,000 km/s) in the absence of matter, or in other words, in a vacuum. (Electromagnetic waves move slower when propagating through matter.) This is astonishingly fast, but light is not instantaneous—it takes some amount of time to travel from one point to the next. What was troubling about Maxwell's equations is that they indicated that the speed of light does not change (except when light moves from one material to another).

Fixed speeds do not make much sense because in everyday activities, speeds are relative to the observer. For example, suppose a dog walks at 3 miles/hour (4.8 km/h) down the aisle of a train traveling at 50 miles/hour (80 km/h). To a person sitting on the train, the dog moves at 3 miles/hour (4.8 km/h) past his or her seat. But to a person standing at a station who watches the train go past, the dog would seem to move at 53 miles/hour (84.8 km/h)—the dog's walking speed adds to the train's speed to give the rate at which the dog moves past the stationary observer. Yet Maxwell's theory suggested light would not behave in this manner. If someone shined a flashlight on a moving train, an observer sitting on the train and a stationary observer at the station would see the flashlight beam move at the same speed. The figure illustrates this point.

Another troubling aspect of Maxwell's electromagnetic waves is that they seemed to propagate without anything to carry them. Waves are disturbances propagating through some kind of medium—sound, for instance, is a wave carried by the motion of air molecules, and ocean waves move along the water. People felt that electromagnetic waves required a substance to carry their motion, and they called this substance the ether. The ether presumably filled all space, and electromagnetic waves were assumed to be oscillations in this substance. To explain the fixed speed of

The train moves at velocity, *v*, and a passenger shines a flashlight toward the front. The passenger observes light moving at velocity *c*. A stationary observer might expect to see light moving at $c + v$ since the train is moving at velocity, *v*, with respect to the stationary observer, and light is moving at velocity, *c*, with respect to the train. Although velocities appear to add in most such situations, it does not happen here.

electromagnetic waves indicated by Maxwell's theory, many scientists believed that this fixed speed was to be measured relative to the ether. In the case of a flashlight on a train, as described above, an observer sitting on the train and a stationary observer would measure a different speed for the beam in the same way as they would for the walking dog. Only the speed relative to the ether was fixed.

The ether idea, while it made sense in view of everyday physics, is wrong. The physicist who challenged this mistaken idea and found the correct solution was Albert Einstein (1879–1955), a brilliant thinker whose creativity, combined with scientific accuracy, made a lasting impact. Previous chapters have already discussed some of Einstein's work, and this chapter examines relativity—ideas centered around Einstein's concept of observers, relative motion, and the laws of physics. These ideas changed everything from perceptions of physics to the view of the universe.

Albert Einstein's Postulates

In an effort to prove the existence of the ether, scientists wanted to do an experiment similar to the flashlight and train example above. But since light travels so quickly, motion faster than a train was necessary; otherwise, the differences measured by different observers would be small and undetectable. Albert Michelson (1852–1931) and Edward Morley (1838–1923) conducted one of the most famous tests of the ether theory in 1887 by using Earth's motion. The sidebar on pages 92–94 describes their experimental procedure.

Michelson and Morley failed to find a change in the speed of light arising from Earth's motion through the ether. One possible explanation for this failure is that the speed of light is the same in all cases. This explanation would be equivalent to admitting that Maxwell's theory needed no ether, so the speed of light would be identical for all observers. Few people were willing to entertain this unusual notion even after the Michelson and Morley experiment, with the exception of Albert Einstein. Einstein had obtained a degree in physics in 1901, but soon afterward he took a job in the patent office of Switzerland, being unable to obtain a job teaching

This interferometer, used by Edward Morley and a colleague, closely resembles the instrument with which Albert Michelson and Morley performed their experiments. *(AIP Emilio Segrè Visual Archives)*

physics (apparently because at least one of his professors did not like him and had written negative things to prospective employers). In 1905, Einstein published a paper outlining a theory that became known as the special theory of relativity.

Einstein began with an extremely important postulate, or assumption, to discard the ether. Maxwell's equations of electromagnetism indicated that light moved at a fixed speed, which as discussed above gave rise to the idea of a fixed ether. Physicists at the time thought that this ether provided a standard frame of reference by which motion, such as the speed of a light wave, could be measured. A frame of reference is like a set of axes extending in space, as illustrated in the figure. A person can measure speed in a frame of reference by measuring the distance an object travels—say, from point *A* to point *B*—and dividing by the time it takes. The ether was supposed to be an absolute frame of reference, and only objects that moved relative to the

The Michelson and Morley Experiment

Michelson and Morley used Earth's motion because the planet travels quickly enough to offer a chance to detect the effects of the ether, if it existed. The radius, *R*, of Earth's orbit averages about 93,000,000 miles (149,000,000 km), so Earth travels roughly $2\pi R$ = 584,000,000 miles (934,000,000 km) in one full orbit. (This calculation approximates Earth's orbit as a circle and uses the formula for the circumference, $2\pi R$. Even though Earth's orbit is slightly elliptical, the calculation is not too far off.) Earth travels this distance in one year, approximately 365.25 days or 8,694 hours. The average velocity is therefore 584,000,000 miles (934,000,000 km) divided by 8,694 hours = 67,000 miles/hr (107,000 km/h). This is only about 0.01 percent of the speed of light, but it is a measurable fraction.

The experimenters compared the speed of light in the direction of the planet's motion through the presumed ether to the speed of light at right angles to this motion. The figure shows a diagram of the experiment. A partially silvered mirror, *M*, reflects part of a light beam to a fully silvered mirror, M_1, and allows the rest of the beam to pass through to a fully silvered mirror, M_2. According to the ether theory, Earth's motion through the ether would add to the speed of light when light traveled in the same direction and would subtract when light traveled in the opposite direction. No one knew what orientation the solar system and its planets had with respect to the ether, so Michelson and Morley conducted their experiment at different positions and times.

M, the partially silvered mirror, acted as a beam-splitter, allowing some of the light from the source to pass, and reflecting the remainder of the beam. If the path to mirror M_2 is parallel to Earth's speed, *v*, as shown in the figure, the path to mirror M_1 is perpendicular. The speed of the M_1 beam will be unaffected, but the M_2 beam should change speed if the speed of light is not constant—the M_2 beam would travel at $c - v$ in one direction and $c + v$ in the other. If this was true, the average speed for the M_2 beam would be less than *c*. Note that the average speed is not found by adding the two speeds and dividing by 2, which would give an incorrect answer of *c*. The reason this simple formula does not work is that the light beam traveled in one direction for a longer period of time than the other. The distance is the same, so the beam would have

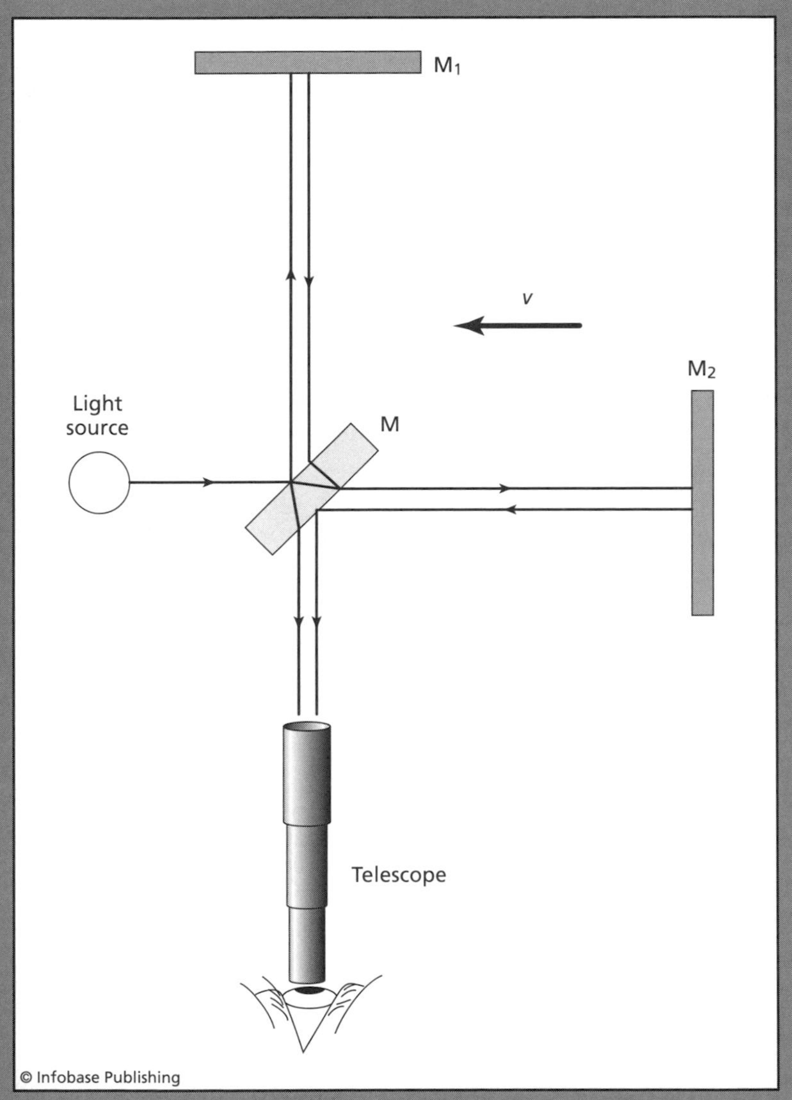

The experimental equipment used by Michelson and Morley included a source of light; a partially silvered mirror,* M*, that transmitted some of the light and reflected the rest; two reflecting mirrors,* M_1 *and* M_2*, along two paths at right angles; and a point of observation to examine the combined light beams.

(continued on next page)

(continued from previous page)
traveled for a longer time at the slower speed $c - v$ than $c + v$. Its average speed is therefore $< c$.

This experimental configuration produces a sensitive measuring instrument called an interferometer. An interferometer measures differences in the speed or path of light by using a wave property called interference. In the experiment, one light beam (according to the ether theory) would slow down and then would speed up. This occurred because Earth's speed subtracted from and then added to the light speed as the beam traveled first against and then with Earth's motion. This is the beam that travels to and from mirror M_2 in the figure. The two light beams combine afterward, and combined waves interfere with one another because, for example, one wave may be at a crest at the same time as the other wave is at a trough, in which case the waves cancel. The different trips taken by the two light beams in the interferometer will cause some of the waves to cancel, causing darkness, and some to add together (they add when both waves are at a crest, so the combined wave is larger), creating brighter light. The result is a set of bright and dark bands called interference fringes.

Michelson and Morley looked for changes in the interference fringes as they rotated their interferometer to different positions. If the speed of one of the beams had changed, the interference fringes would be different. But the experimenters found no such differences. The ether, if it existed, did not have the expected effect.

ether were considered to be truly in motion, since the ether was the only thing that was absolutely at rest. According to the ether theory, the ether was the one and only frame of reference in which Maxwell's equations would apply, for it was the only frame of reference in which light had a fixed speed. For any observer who was in motion relative to the ether, Maxwell's equations required modification.

Impressed with Maxwell's theory, Einstein believed that the equations Maxwell had discovered should be the same for all reference frames. It did not make sense to Einstein that the laws of physics had to change simply because a person made measurements using a different coordinate system. In a postulate called the principle of relativity, Einstein stated that the laws of physics are the same for all reference frames moving at a constant speed. The requirement for constant speed will become apparent in a later section, which discusses changes in speed (acceleration) and gravitation. The 1905 theory only applied to this special situation (constant speed) and is known as the special theory of relativity.

Einstein made another postulate, declaring that the speed of light was the same for all observers. This postulate simply reaffirms what Maxwell's equations suggested. Armed with only the

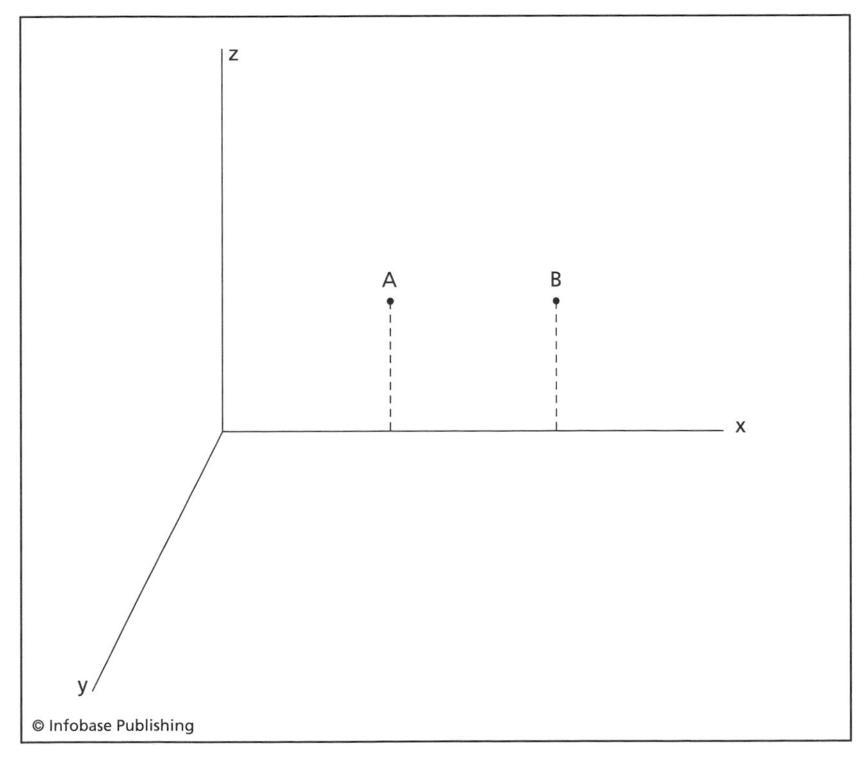

Axes *x*, *y*, and *z* form a coordinate system or frame of reference. Each axis is perpendicular to the other two (in other words, each two axes make a 90-degree angle), with the *x*- and *z*-axis lying in the plane formed by the paper and *y* protruding out of the plane. (The *y*-axis is drawn slightly bent so that it can be seen.) Each point, such as *A* or *B*, can be represented by its position relative to each axis. The figure shows the position along the *x*-axis for both *A* and *B*.

principle of relativity and his convictions about the speed of light, Einstein proceeded to deduce some startling consequences.

Time Dilation and Length Contraction

Events occur at specific locations, and people describe those locations using a frame of reference (coordinate system), such as three perpendicular axes show in the previous figure. Three numbers called spatial coordinates represent distances from the three axes and identify a point in space. This concept is old, originating centuries ago when French philosopher and scientist René Descartes (1596–1650) developed it. What Einstein realized is that time is just as important as the three spatial coordinates. In Einstein's system, events must be specified by three spatial coordinates and one time "coordinate." This system is known as space-time.

Time is critical because the time of an event depends on the observer's frame of reference. To understand this, consider the situation diagrammed in the figure on page 97. Suppose two lightning bolts strike the ends of a train that is traveling at a constant velocity, v. The bolts mark the train at points A' ("A prime") and B' and also leave marks directly underneath, on the ground, at points A and B, as shown in the figure. One observer is standing on the train halfway between the two ends, at point P', and another observer is standing on the ground at point P, directly below point P'. An important question to ask is whether the two lightning bolts occurred at the same time. The strange thing is that the observers will not agree on the answer!

Suppose that the stationary observer at P claims the lightning bolts struck simultaneously—at the same time. This means that in the frame of reference of P, light from a bolt traveled the equal distance from A to P and from B to P and arrived at the same time. But the observer on the train at P' has a different frame of reference—the moving train. This observer is moving toward B at constant velocity, v, and away from point A at the same velocity. Light travels at the same speed for all observers, so light from the bolt at the front of the train will arrive at P' sooner than light from

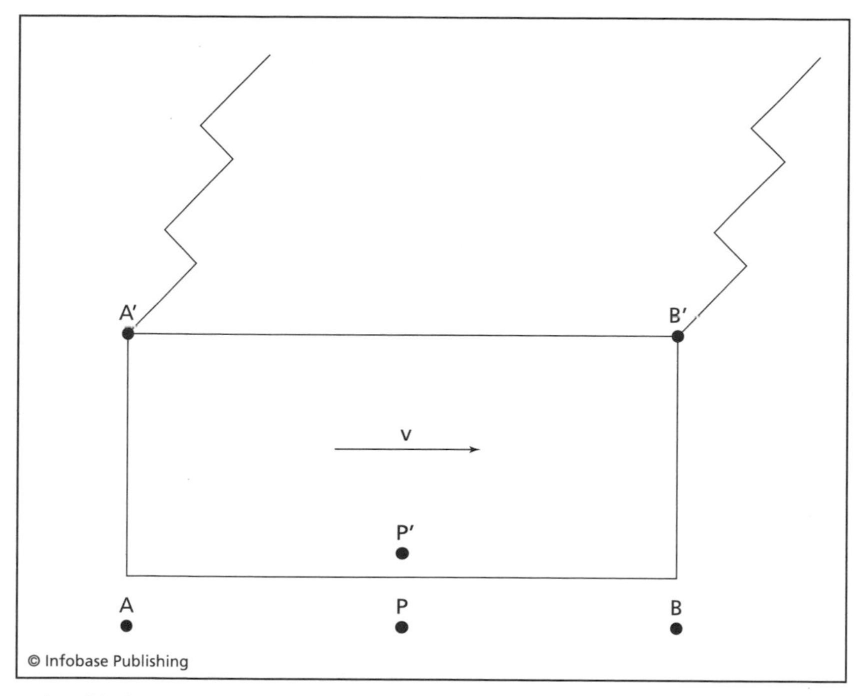

Bolts of lightning strike the train (shown as a box) at the front, *B'*, and rear, *A'*, as observed by a person, *P'*, sitting in the middle of the train. As the bolts strike the train, they leave marks on the ground at *B* and *A*, underneath the train, as observed by a stationary person, *P*, standing halfway between these two points, directly underneath the position of *P'*.

the rear because it has less distance to travel—as the light is traveling, *P'* has moved toward the front lightning bolt and away from the rear one with a speed of v, the speed of the train. According to the point of view of the observer at *P'*, the lightning bolts did not strike at the same time. The observer at *P'* reports that the lightning bolt at the front of the train struck first.

Einstein believed that each observer is correct in his or her own frame of reference. The principle of relativity requires this to be true, for if the laws of physics are the same for all frames of reference, each observer must be able to make correct measurements. As a result—and although it sounds quite strange—the time an event occurs is not the same for all people, but instead depends on the observer's frame of reference. Time is relative.

The Equations of Special Relativity

Einstein derived these equations based on the mathematics and geometry of situations such as the moving train, as described in this chapter. The derivation of the equations will not be covered here, but it is interesting to see the equations and plug in a few numbers.

If the velocity of the observer in motion is v, the following equation compares the time interval Δt of the stationary observer with the time interval $\Delta t'$ of the moving observer:

$$\Delta t = \frac{\Delta t'}{\sqrt{1 - \frac{v^2}{c^2}}}.$$

where c is the speed of light in a vacuum and $\sqrt{}$ is the square root symbol. (The Greek letter Δ, delta, represents a change in a quantity, such as the change in time associated with an interval.)

The denominator is a number greater than 0 but less than or equal to 1. If $v = 0$, the denominator is 1; v must be less than c since no object other than a photon can travel as fast as light in a vacuum, so the denominator is always greater than 0. Dividing by a number less than 1 increases the magnitude, so Δt is either larger than $\Delta t'$ or, if $v = 0$, it is the same. This means that the time interval measured by an observer in constant motion ($v > 0$) is longer, or dilated, when measured by a stationary observer.

The consequences of a lack of agreement between observers in different frames of reference are so odd that some people have trouble believing them at first. As shown in the example above, even the simple decision of simultaneity is not the same between observers in different frames of reference. Both space and time are involved, and observers moving relative to one another do not measure the same quantities of length and time corresponding to the same object or event. To a stationary observer, the clock of an observer in motion seems to run slower and the yardstick seems to shrink. These effects are known as time dilation (a slowing of time) and length contrac-

Time dilation does not seem to occur at velocities that people normally encounter. It actually does occur, but the effect is so small that it is virtually impossible to notice. Consider the lightning and train example, where v is 50 miles/hour (80 km/h), which equals 0.0139 miles/sec (0.0222 km/s). Using this velocity along with c = 186,200 miles/second (300,000 km/s), the denominator of the equation is $\sqrt{[1 - (0.0139)^2/(186{,}200)^2]} = \sqrt{(1 - 5.6 \times 10^{-15})} = 0.999999999997$. The time interval is different, but only by a few trillionths or so. An interval of a minute would be different by a tiny and undetectable fraction of a second.

But suppose v = 93,100 miles/second (150,000 km/s), which is one-half the speed of light in a vacuum. Then the denominator is $\sqrt{[1 - (93{,}100)^2/(186{,}200)^2]} = \sqrt{(1 - 0.25)} = 0.866$. An interval of a minute as measured by an observer moving at v = 93,100 miles/second (150,000 km/s) would be longer to a stationary observer by more than 9 seconds (60 seconds/0.866 = 69.3 seconds).

The formula describing length contraction is similar. A length, L', as measured by an observer moving at velocity, v, would equal a length, L, measured by a stationary observer given by the following equation:

$$L = L'\sqrt{1 - \frac{v^2}{c^2}}.$$

Plugging in some numbers shows that length contraction is also not noticeable except when the velocities are high enough to be a considerable fraction of c.

tion. The sidebar describes the equations governing how much time dilates and length contracts.

The effects of special relativity are so unusual because they are not encountered in everyday activity. The formulas for time dilation and length contraction, given in the sidebar, involve the square of c, the speed of light. Since c is such a huge velocity, it overwhelms the much slower speeds to which people are accustomed. Time dilation and length contraction are detectable only when the motion between observers is significant when compared to c. Because c is 186,200 miles/second (300,000 km/s), or about 670,320,000 miles/hour (1,080,000,000 km/h), it would require remarkable precision to detect time dilation or length contraction due to the motion of a train or automobile.

The principle of relativity—the laws of physics are the same for all observers—was Einstein's motivation for his theory, but relativity also required a slight change in the formulas derived in the 17th century by Sir Isaac Newton. The Newtonian formula for a particle's momentum is the product of its mass, m, and its velocity, v, and this quantity, mv, is important in physics because it is conserved—the momentum before and after a force acts or a collision occurs is the same. But in relativity theory, the momentum of a particle with mass, m, and velocity, v, is given by the following formula:

$$\frac{mv}{\sqrt{1-\frac{v^2}{c^2}}}.$$

When v is small, the denominator is very close to 1, so the old formula is approximately correct. Only for high speeds, comparable to the speed of light, does relativity's formula differ significantly from the old one.

Sometimes people interpret relativity's momentum formula to also be the product of mass and velocity, but in relativity the "mass" of a particle would be given by

$$\frac{m}{\sqrt{1-\frac{v^2}{c^2}}}.$$

This quantity is known as relativistic mass, and it increases with increasing velocity. The formula shows why trying to accelerate an object or a particle that has mass up to (or past) the speed of light is futile. As v increases, so does the mass, and the mass becomes so great as v approaches c that no amount of energy can cause it to reach the speed of light. No object or particle that has any mass can ever be accelerated up to light's speed. Even the gigantic accelerators discussed in the previous chapter cannot force a tiny mass such as an electron or a proton up to the speed of light, much less beyond.

Although some people have difficulty believing time dilation, length contraction, and "relativistic mass," plenty of experimental evidence supports the special theory of relativity. Particles in the accelerators discussed in the previous chapter routinely reach speeds close to *c*, where relativistic effects become detectable, and the behavior of these particles is always in accordance with the theory.

Time dilation is also a measurable effect. Another way of expressing time dilation is to say that clocks run slow when they are in motion, meaning that they run slower than a stationary clock does. Time dilation affects all processes, from the ticking of a clock to the aging of a person. A person traveling at high speed does not grow as old as quickly as a stationary one, and while this effect has not been observed directly—no machine exists that can move a person at a considerable fraction of the speed of light—a similar effect occurs for a particle called the muon. These particles are unstable and have short lifetimes, quickly decaying. But researchers at CERN used their accelerators to give muons a velocity extremely close to *c*, the speed of light, and found that the average lifetime of these particles was about 30 times greater than that of stationary muons. Humans in the same situation would have an average life expectancy of more than 2,000 years, as measured by stationary observers.

The Twin Paradox

Although relativistic time dilation is true, it does not mean that time appears to be slower to the people who are traveling at high speed. Sometimes people mistakenly envision activity taking place in a moving frame of reference as being in "slow motion." But a slowing of time would only be noticed by a stationary observer watching the people who are in the moving frame of reference. To a person in motion, everything seems normal, including time. The principle of relativity demands that this be so, otherwise the laws of physics would be different in different frames of reference. Special relativity's effects occur only for relative motion, not within a single frame of reference.

But there is a situation in special relativity that seems to be a paradox—a contradiction that cannot be true. Suppose that Albert, an astronaut, embarks on a space journey at a rocket traveling at close to the speed of light. (Although present-day rockets cannot achieve such speeds, there is no reason not to believe people will eventually be able to build one.) Albert leaves behind his twin brother, Balbert, who stays on Earth. According to the special theory of relativity, Albert will age less than Balbert, and when he returns, he will be younger than his twin.

Sometimes the fact that Albert will be younger than his twin is stated as a paradox, but there is no paradox here. Because of time dilation, the clocks in Albert's spaceship, including his biological clocks, tick more slowly than Balbert's. When Albert returns, he will be younger than his twin.

The paradox is that motion is relative. It is impossible to distinguish between frames of reference moving at a constant rate of speed—this is the principle of relativity. An observer on a train moving at a steady 50 miles/hour (80 km/h) watches the scenery go past the window at the same speed. To the observer on a train, a tree seems to rush by at a steady 50 miles/hour (80 km/h). Although people know that trains move and trees do not, as far as physics is concerned, there is no distinction in the two situations. If for some reason a person got on a train and then the scenery started to move at a constant speed—perhaps this could occur in an amusement park ride—there would be no experiment the observer on the train could perform to reveal that the scenery is in motion and not the train. The laws of physics would be the same in both cases.

In Albert's point of view, he is traveling at close to the speed of light. But suppose Balbert says that he is actually the one moving instead of Albert. Balbert justifies his point of view with the principle of relativity. If Balbert wants to claim that he is the one in motion and that his brother Albert is standing still, he has a right to do so. There is undeniable motion between the two brothers, but which one is moving depends on whose point of view is adopted. Since Balbert has a right to claim that he is the one in motion, his clocks should be running slower than Albert's. As a result, he should be younger than his twin brother when Albert returns. Yet Albert is of the opinion that he will be younger than Balbert. This is known as the twin paradox. Both cannot be right.

If the predictions of a theory create a paradox, it must contain a flaw. But there is no paradox created by the special theory of relativity because the flaw is in Balbert's argument. Albert was in the same frame of reference as his brother when he began his journey, and he needed the acceleration of a powerful rocket to bring him up to speed. Acceleration involves a change in speed. Observers cannot determine who is moving and who is not when the relative motion is at constant speed, but they can definitely distinguish between frames of reference when there is a change in speed. Observers in an accelerating rocket or in any other accelerating vehicle feel as if their seat is pushing against their back. Albert is the one who experienced the acceleration, so Albert is the one who sped up. Special relativity is not valid during changes in speed, but the theory remains valid for Balbert, who will observe Albert's clocks running slower. When Albert returns, he will be younger than his twin brother.

An observer on a train feels a push in the back as the train speeds up, so this effect also ruins the train experiment mentioned above. An observer who boards a train and then sees the scenery start to move will know that something is wrong because there was no acceleration—no push in the back. Once the train is running at a constant speed, there is no way to distinguish whether the train or the scenery is in motion, but this is not true while a frame of reference is slowing down or speeding up.

Gravity and the General Theory of Relativity

Einstein noticed the problem with acceleration, and he tried to adapt the special theory of relativity to more general situations, including changes in speed. Although this attempt was not successful, Einstein discovered another principle, called the principle of equivalence, relating acceleration and gravitation. In its simplest form, the principle of equivalence states that there is no distinction between uniform acceleration—changing speed at a constant rate of change—and the action of the force of gravitation. Einstein used this principle to extend relativity theory to more general cases. This theory is the general theory of relativity, but it is also the best explanation of gravitation that physicists have today, so it is a theory of gravitation as well.

As mentioned above, an observer on a train moving at constant speed would be unable to perform an experiment that could reveal whether the train was moving or the scenery—there is no law of physics that would be different if the train were moving as opposed to the scenery, so no experiment could tell which one applies. Acceleration spoils this effect, though, because changes in speed are detectable.

But imagine another situation. Instead of a train, consider an observer, Alice, standing in a box with no windows, like an elevator. Suppose Alice initially has no idea where the box is located. It can be anywhere, in space or on a planet. In order to gather clues, she makes some observations and performs some experiments.

Alice notices that she has weight—she must engage her leg muscles in order to keep from falling down. When she pulls out a coin and drops it, the coin falls to the floor. Measuring the speed of the coin as it falls, Alice determines that it accelerates at a constant rate—the coin accelerates at 32 feet/second^2 (9.8 m/s^2). In other words, the coin's speed increases at a constant rate of 32 feet/second (9.8 m/s) for every second it falls. After one second its speed is 32 feet/second (9.8 m/s), and after two seconds its speed is 64 feet/second (19.6 m/s). Although Alice cannot see outside and has no other clue regarding the location, her conclusion, based on the experimental evidence, is that she is on Earth's surface. She has weight, which she assumes is due to gravity, and the coin behaves exactly as it would if dropped on Earth's surface, accelerating at 32 feet/second^2 (9.8 m/s^2).

Although Alice's observations are valid and intelligent, Einstein realized that her conclusion could be wrong. Instead of on Earth's surface, suppose the box is in space and accelerating at a constant rate of 32 feet/second^2 (9.8 m/s^2). Acceleration, as noticed by the observer on the train, feels like a push. This is the same as having weight, for if the box was accelerating upward, Alice would feel a push from the floor that would feel the same as "weight." Astronauts experience this same effect during the extreme accelerations of rocket launches and even describe these accelerations in terms of gravity—1 "G," or 1 gravity of acceleration, is an acceleration of 32 feet/second^2 (9.8 m/s^2).

The coin's acceleration would be explained if the box is accelerating at 32 feet/second^2 (9.8 m/s^2). This would be 1 G. Another way of looking at this situation is to consider the box's motion. When Alice drops the coin, its initial velocity is the same as hers and that of the rest of the box. The box continues to accelerate, increasing its speed at a rate of 32 feet/second^2 (9.8 m/s^2), but the coin no longer feels this push because it is no longer firmly attached to the box or something in the box—it is freely moving, so it continues to travel at its initial velocity. As a result, the floor of the box approaches or "gains" on the coin as the box increases its velocity because it picks up speed as it accelerates. The coin appears to "fall" to the floor, accelerating at 32 feet/second^2 (9.8 m/s^2), although it is actually the box that accelerates.

The principle of equivalence states that Alice cannot do any experiment to determine whether she is experiencing the force of gravitation or the effects of a constant rate of acceleration. The laws of physics are the same in both situations. This result—that scientific laws do not depend on the observer's location or motion—is the great foundation on which both the special and general theories of relativity rest.

Einstein also realized that the principle of equivalence says something about gravitation and mass. If an observer cannot tell the difference between weight due to gravity and the push of acceleration, then mass in these two cases must be identical. Sir Isaac Newton formulated a theory some 300 years ago that related acceleration to the concept of a force—a push or pull—in which he said forces cause an object to accelerate at a rate that depends on the object's mass. Another one of Newton's achievements was his discovery of an equation for the force of gravitation, which involves an object's mass. Acceleration and gravitation both involve mass, and if the principle of equivalence is true, the value of an object's mass should be the same in both cases. Measurements indicate that this is true.

Newton's equations for motion and gravitation worked well—at least for speeds much less than the speed of light, as described above—but the notion of a force bothered Einstein. To Newton, objects with mass exert an attractive gravitational force on one another, even if the objects are distant. Einstein did not understand how objects could reach out across a distance and exert forces. The Standard Model of particle physics, explained in the previous chapter, regards a force as something created by the exchange of force-carrying particles, but this theory had not been developed in the early 20th century, when Einstein was pondering the nature of forces. (It is also questionable whether Einstein would have liked this theory.)

Einstein thought in terms of geometry rather than forces. Acceleration and gravity are indistinguishable, so an object's gravitational attraction to another would be equivalent to a falling or rolling toward it. Einstein viewed gravitation as a warping of space-time itself, a curving of its geometry. A massive object such as the Sun curves space, and objects such as planets follow these curves. Physicists sometimes view space-time as if it were a rubber sheet,

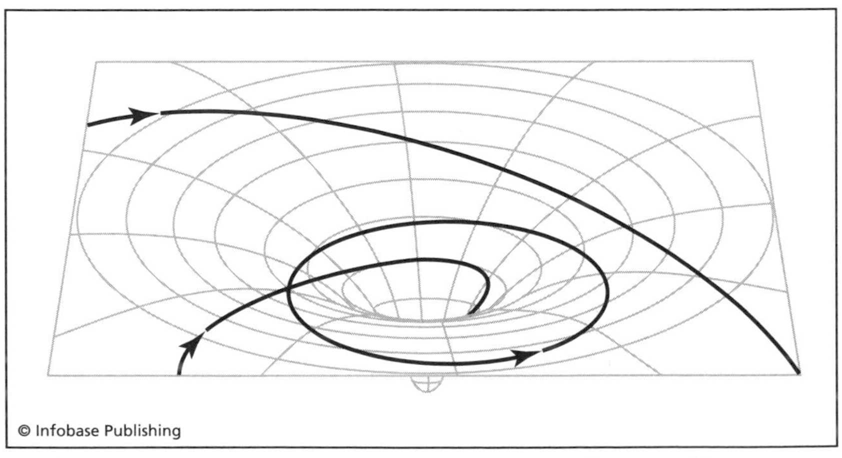

This figure depicts a massive object as a steep dent or valley in a rubber sheet. The rubber sheet represents space-time, and the curve of the valley represents gravity. Other bodies roll toward the valley—this is the geometrical representation or model of the manner by which the mass of the object attracts other bodies.

and massive objects resting on it create indentions that curve the sheet, as shown in the figure. As other objects approach, they fall toward the massive body.

Einstein proposed the general theory of relativity in 1915. By this time he was no longer an unknown patent clerk but an acclaimed physicist, already famous from his papers published in 1905. Einstein received the Nobel Prize in physics in 1921.

But physicists needed to test Einstein's theory before they accepted it. The test for equivalence of mass under gravitation and acceleration was mentioned above, but the general theory of relativity made several other key predictions. Einstein derived the general theory's equations by the same kind of reasoning process that he used for the special theory, except for the general theory he considered situations such as Alice's—what consequences would follow, Einstein wondered, from the principle of equivalence. The theory altered formulas derived by Newton, although as was the case with the special theory of relativity, the new formulas reduced to the old ones in many situations. But physicists could test the differences.

One of the predictions of general relativity could be confirmed right away. The orbit of the planet Mercury shows a small but noticeable shift in its perihelion (the closest approach to the Sun)—the perihelion changes slightly over the years, a fact known since the 19th century. Gravitational effects from other planets accounted for a portion of the shift, but not all. The equations of general relativity explained the rest of the shift, clearing up the long-standing astronomical mystery.

General relativity also predicts the bending of light due to gravitation. The older theory of gravitation, based on Newton's idea, predicts this phenomenon as well, but general relativity indicates that light bends two times more than would be expected from the old theory. The reason for the increased bending is the curvature of space-time; according to Einstein, massive objects warp the space around them, and all trajectories, including those of light beams, are bent. This effect has been confirmed in a number of ways, including measuring the change in position of stars as their light passes close to the Sun. The change in a star's position is due to the deflection of starlight by the presence of the Sun, and the results agree well with general relativity.

Another prediction of general relativity is even more interesting, though quite similar to the time-dilation effect of the special theory of relativity. The special theory correctly predicts the slowing of time due to motion, and the general theory of relativity predicts the slowing of time due to gravitation. This prediction, like that of special relativity, has been confirmed. In one of the most sensitive tests, called Gravity Probe A and conducted in 1976, a rocket lifted an extremely accurate clock to an altitude of 6,200 miles (10,000 km). At this height the clock should run slightly faster than a clock at Earth's surface since the force of gravitation is not as strong at this altitude. The effect is tiny—only a few billionths of a second per minute—but the experimental results confirmed the prediction.

An even more accurate experiment, called Gravity Probe B, put a small satellite into orbit in 2004. This probe will measure the effects of Earth's gravity on space and time. The results are not yet in, but physicists expect an answer to emerge within a few years. Meanwhile, general relativity as well as special relativity have proven their importance to other satellites, such as the satellites responsible for signals used in the Global Positioning System (GPS). This system provides navigators with highly accurate information on their position—with a GPS receiver, a ship's captain, for example, can know the exact location of his ship on the ocean, in some cases to within less than 10 feet (3 m). GPS works because orbiting satellites emit signals providing the time of emission and their position. A GPS receiver detects these signals and uses them to compute its own position. The extreme accuracy of GPS requires remarkable precision of the satellite signals, which cannot be obtained unless the high-speed, high-flying satellites take relativity theory into account.

General relativity is a complex theory. One of its predictions that has yet to be confirmed is the existence of gravitational waves. These waves are similar to the electromagnetic waves of light. According to the theory, objects such as masses that orbit large stars will emit these gravitational waves, also known as gravita-

Developed by NASA, Stanford University, and Lockheed Martin, the mission of Gravity Probe B, shown here at Vandenberg Air Force Base in California, is to test Einstein's general theory of relativity. *(NASA-KSC)*

tional "radiation." Gravitational waves should not be confused with the graviton, a particle predicted by particle-physics theory but not yet found, as discussed in the previous chapter. But if both exist, they may be two realizations of the same phenomenon, as are photons—particles of light—and electromagnetic waves.

No one has yet detected a gravitational wave, probably because its magnitude is tiny, as predicted by the theory. The Laser Interferometer Gravitational-Wave Observatory (LIGO), a project conducted by numerous scientists, uses interferometers as detectors. Interferometers, as discussed above, detect changes based on wave interference. The passing of gravitational waves would distort lengths, compressing and extending space-time as they ripple past. The challenge is to detect these ripples, which physicists expect would normally be less than the size of a proton in most cases.

The existence of gravitational waves, if they are ever confirmed, would be important to astronomers because these waves would give them another tool with which to study the universe. Gravitational-wave astronomy, if it ever comes about, could yield exciting new clues about the nature of the universe, just as the study of electromagnetic waves such as light, radio waves, and X-rays has been a rich source of information. Many people expect gravitational waves will be found, because some of the most celebrated astronomical predictions of general relativity have already been confirmed, or at least are widely accepted. One of these phenomena involves the possibility of a strange astronomical object that has become known as a *black hole.*

Black Holes

A black hole is an extremely dense object, containing a huge amount of mass in a tiny volume. This density creates gravity so strong that nothing, not even light, can escape. (There are some unusual exceptions to this rule, based on quantum mechanical phenomena described in chapter 2.) The "hole" is not a hole in space; it is a region where gravity pulls in matter, and this matter will never be seen again.

The equations of the general theory of relativity involve mathematical objects called tensors that are far more complicated than ordinary numbers. The mathematics of general relativity is so difficult that few people have ever mastered it. But shortly after Einstein developed the main ideas of general relativity in 1915, German physicist Karl Schwarzschild (1873–1916) found a solution to the equations that described a black hole, although the term *black hole* did not appear until the 1960s.

A black hole may form when an old and massive star collapses. A star is a huge sphere of gas, mostly hydrogen and helium, and it emits light from nuclear fusion reactions as described in chapter 1. The force of gravitation from the large mass pulls the star's matter inward, just as objects fall on Earth's surface due to gravity. Countering this inward pull is an outward pressure due to the energy released from the nuclear reactions. Throughout most of the star's lifetime, the inward and outward forces balance and the star maintains a steady radius. But when the star begins to exhaust the material it needs for fusion, gravity's inward pull exceeds the outward pressure and the star collapses.

For a small star, such as the Sun, the collapse at the end of its lifetime does not involve as great an amount of matter and will not generate a black hole. A black hole forms only when there

The giant star in this illustration would probably end up as a black hole. Underneath the star is a diagram of the solar system of the Sun. The radius of the giant star stretches well beyond Earth's orbit. *(NASA/JPL-Caltech)*

is so much mass that its particles cannot withstand the resulting gravitational attraction. This may occur for stars with perhaps 10 times more mass than the Sun. (Smaller stars such as the Sun also have longer lifetimes, lasting billions of years. The Sun has several billion more years before it exhausts its nuclear fuel.) A collapse of a massive star compresses its matter into a black hole, at least in theory.

The outer boundary of a black hole is known as the event horizon. Once an object passes this "horizon," it cannot escape. Black holes have boundaries, or diameters, that are incredibly small for the amount of mass they contain. A black hole with 10 times the mass of the Sun would have a diameter of about 37 miles (59 km). The Sun, which has only a tenth of this mass, has a diameter of 875,000 miles (1,400,000 km).

No one knows what happens to matter inside a black hole. The center is a region called a singularity, a point where the density of matter cannot be mathematically described because, according to the equations, it is infinite. But the mass of a black hole continues to exert gravitational attraction. Gravity is how such objects can be located.

The strong gravitational attraction of a black hole tugs on any nearby particles. This powerful force accelerates the particles at a much higher rate than Earth's gravity, and these pieces of matter collide and become hot. Matter falling into a black hole often has enough energy to radiate high-frequency electromagnetic radiation called X-rays. Another effect of strong gravitational fields is a significant bending of light, as predicted by general relativity. This bending can become so great that it produces a lens effect, as light from a more distant object passes through this region. As a result of the bending, the image and position of the distant object may be distorted. Astronomers have found hundreds of objects that may be, and in some cases probably are, black holes. The center of the Milky Way galaxy, the galaxy that contains the Sun and its solar system, appears to harbor a black hole with a mass exceeding one million times that of the Sun.

Many people have wondered what it would be like to explore the inside of a black hole. The result would be unfortunate, for not only would the intrepid explorers fail to return, but also the strong gravitational forces would kill them. A black hole's gravity would be deadly because its strength changes with distance, and this would result in the stretching of an explorer's body. Earth's gravity also changes by the same rate, but this is not a problem for the planet because Earth's gravity is not as strong as that of a black hole. Near the singularity of a black hole, gravity is so intense that there is a significant change in its strength over small distances. The gravitational force on one end of the explorer's body would be

This jet of particles comes from extremely hot matter swirling around a massive black hole in the center of galaxy M87. The mass of this black hole is several billion times that of the Sun. *(NASA/STScI/AURA)*

far greater than on the other, elongating the body—an unpleasant way to die.

The mathematical richness of general relativity allows the existence of other objects as strange as or stranger than black holes. A wormhole is a passageway connecting two regions of space-time, and perhaps in some cases the regions may be quite distant in terms of space or time. Some people have attempted to relate wormholes to the interiors of black holes. Although such ideas have sparked the imagination of a small number of scientists and an even larger number of writers, no evidence for these objects exists.

Even if wormholes are never discovered, relativity theory has had an enormous impact on physics. Both the special and general theory of relativity modified long-held notions and formulas and altered how people perceive space and the universe. Physicists continue to investigate the mathematical complexities of the general theory of relativity, in part because of its richness and in part because it remains the best theory of gravitation available today.

Although gravitation is not important to the particle physics discussed in the previous chapter, this force is extremely important over the vast distances encountered throughout the universe. Because of this, the general theory of relativity is vital to physicists and astronomers who study the universe. General relativity's relevance to astronomers became clear quite early on, as Einstein himself discovered that his equations predicted the universe was increasing in size. Believing such an expansion was unlikely or impossible, Einstein modified the equations so that this result was no longer valid. The modification proved to be a mistake, and as the following chapter discusses, general relativity yielded more insight into the universe than even its discoverer could have believed.

5

COSMOLOGY

AFTER THE 17th century, when Sir Isaac Newton discovered equations relating the motion of an object and the forces acting on it, people gained a better understanding of the world than they ever had before. Physicists could calculate and predict the behavior of objects in all kinds of situations, whether in motion or at rest. Newton's theory of gravitation extended the reach of science all the way throughout the solar system and beyond, explaining the orbits of planets and allowing them to be calculated with great accuracy. Everything seemed to obey these laws of physics, and people imagined the universe was like a gigantic but simple mechanical device, operating forever and never changing.

Most people assumed this static (unchanging) universe was infinite in extent—it had no boundaries. But something puzzled German scientist and physician Heinrich Olbers (1758–1840). If the universe was infinite and filled with stars, as astronomical observation suggested, there would be an infinite, boundless number of stars. There must be a star at every point in the sky. But if this were the case, the sky would not be dark at night. A dark sky seemed impossible if the universe is infinite, a contradiction that became known as Olbers's paradox.

Although Olbers's paradox raised doubts about an infinite extent of the universe, the idea of a finite (bounded) universe also posed troubling questions. If the universe is not infinite, what kind

More than a million galaxies appear in this grand view of the entire sky. *(NASA-JPL)*

of wall or boundary terminates it? What lies on the other side of the wall? There are no easy answers to these questions. But a valuable clue appeared when Einstein developed the general theory of relativity in the early 20th century, as described in the previous chapter, and physicists began thinking about curved space-time. Then American astronomer Edwin Hubble (1889–1953) showed that the universe was not static after all.

The Big Bang

As mentioned in the previous chapter, the discoverer of the general theory of relativity, Albert Einstein, realized that the theory's equations indicated that the universe is expanding. Einstein at first refused to accept it. Instead, he modified the equations by including a "cosmological constant," which caused the theory to describe a static universe. The term *cosmology* refers to the study of the universe and is derived from the Greek word *kosmos,* meaning *order* or *universe.*

Even though Einstein was a brilliant physicist, several scientists were not convinced that Einstein's modification was necessary or even valid. Based on the complicated equations of general relativity, Belgian scientist Georges Lemaître and Russian mathematician Alexander Friedmann suggested that the prediction of an expanding universe was the correct one. Lemaître later proposed a theory that the universe had been tiny in the past and had grown to its present size after a long time of expansion. Few people believed these notions or even paid much attention until Hubble and other astronomers discovered experimental evidence for the expansion.

Hubble found in 1929 that galaxies in the universe seem to be flying away from Earth in all directions. A galaxy is a vast collection of stars and dust, such as the Milky Way galaxy, which contains the Sun and the solar system along with about 300 billion other stars. The universe contains a huge number of galaxies of a variety of shapes and sizes, and Hubble discovered that their apparent motion carries them away from Earth at a speed depending on their distance—the farthest galaxies recede faster. (A few of the nearest galaxies are moving closer rather than farther away from Earth, due to the gravitational attraction between the Milky Way and these nearby galaxies.) Einstein acknowledged that his modification was an error.

Measurement of this recession of galaxies cannot be direct. Even the closest galaxy is 147,000 trillion miles (236,000 trillion km) away from Earth, and most galaxies are much farther away than that. Hubble had no means of determining the speed of a galaxy except by examining its light. Astronomers study objects by using huge telescopes, which gather and focus the tiny amount of light that travels all the way from distant stars and galaxies to Earth. This light can be spread out into its spectrum, or frequency components, that make up the "colors" of white light, just as Newton used a prism to show that light is composed of the colors of the rainbow.

Hubble could measure the speed of a galaxy from its spectrum because of a wave phenomenon called the Doppler effect, discovered by Austrian physicist Christian Doppler (1803–53). The Doppler effect is a change in the frequency of a wave due to motion between the source and the receiver. The source emits a wave, such as a sound wave, and if there is motion between the source and the receiver—for instance, the siren of a speeding ambulance emits a loud noise, heard by a person standing on the sidewalk—then the frequency as measured by the receiver is not the same as that of the emitted wave. When the source and the receiver are getting closer, the wave increases in frequency—the crests of the wave bunch up—and in the opposite case, the frequency decreases because the distance between crests grows. This effect explains the rise in pitch (frequency) of a siren, as heard by a listener standing on the sidewalk, when the ambulance approaches the listener. The fall in pitch as the ambulance recedes is due to the same effect.

The Doppler effect occurs in light as well as in sound, and it becomes important in astronomy because of lines found in the spectrum of stars and galaxies. German scientist Joseph von Fraunhofer began the study of these lines in the early 1800s when he found them in sunlight. The lines are due to the absorption of specific frequencies of light by atoms—an atom such as hydrogen absorbs only specific frequencies of light, an effect related to its emission spectrum, discussed in chapter 2. Atoms along the path of light, in the outer layers of a star or surrounding a galaxy, for instance, absorb light at their specific frequencies. The absorption causes a loss of light in these frequencies, creating absorption bands or dark lines in the spectrum since the frequencies are not as bright as the others. The absorption and emission frequencies of each atom are known, but motion between the source of light (stars and galaxies) and the receiver (astronomers and their telescopes) changes these frequencies. This is the Doppler effect.

A measure of the speed of motion between a galaxy and Earth can be found in the amount of frequency shift in the absorption and emission lines in the galaxy's spectrum. Hubble discovered that the lines in all the distant galaxies were much lower in frequency, indicating that the galaxies were receding. This shift downward was toward the red end of light's spectrum, the lowest frequency, so the phenomenon is known as *redshift.* Farther galaxies were more redshifted, indicating faster speeds.

The expansion of the universe suggested that Lemaître's idea had merit. For some reason there was an explosion, and evidence based on Hubble's work and observations by other astronomers and physicists indicated that this explosion occurred about 14 billion years ago. This was the birth of the universe, the *big bang.*

This collection of four images shows a distant galaxy that formed very early in the universe's history. Dust obscures its image in visible light (upper right), but it appears in infrared images taken at various wavelengths (upper left, lower left, and lower right). *(NASA-JPL)*

Several aspects of the universe's expansion appear confusing at first. Many physicists prefer to think of space as expanding rather than thinking of the galaxies as being in motion—in this view, the galaxies are not speeding away from Earth but are being carried by an expansion of space, which at least in the present state of affairs overpowers the weak gravitational attraction between distant pieces of matter. Another important point to consider is that the recession of galaxies away from Earth does not mean this planet is the center of the universe. Almost everything in the universe is receding from everything else, like dots drawn on an expanding balloon, illustrated in the figure on page 120. There is no center of the expansion. From the perspective of an astronomer on Earth, all galaxies (except a few of the closest) are receding from this planet, but the same observation would be made by an astronomer on any other planet in the universe.

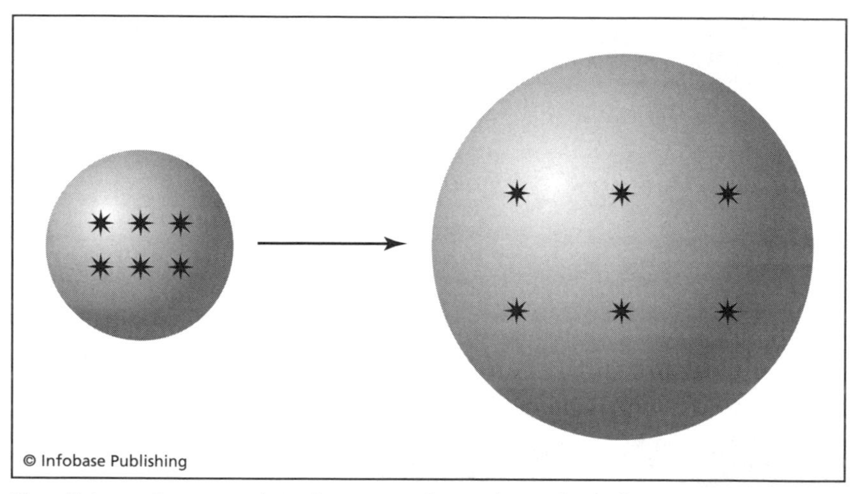

The distance between dots drawn on the surface of a balloon increases as it expands. From the perspective of any one of the dots, all the other dots are receding.

There is widespread acceptance of the big bang theory today, but redshifts are only part of the evidence. The details of the theory are complicated and will not be discussed here, but the theory correctly predicts the amount of hydrogen and helium in the universe. Another key piece of evidence is the existence of the remnants of the explosion that created the universe—an afterglow called the cosmic background radiation.

The cosmic background radiation is electromagnetic radiation coming from all directions in the sky, which is why it is called "background"—it seems to form part of the sky's background. Two scientists, Arno Penzias and Robert Wilson, discovered this radiation by accident in 1964—annoyed by a persistent source of noise at certain frequencies, Penzias and Wilson found that the source was all of space.

Properties of the cosmic background radiation agree with the big bang theory. For instance, the temperature of the cosmic background radiation—and therefore of space—is –454.8°F (–270.4°C), which equals 2.7 K, just a few degrees above the coldest possible temperature. Satellites have recently mapped the radiation in more detail, and these more precise measurements further support the theory. (Note that although it may sound strange to speak of a temperature of space, the cosmic background radiation is energy and could warm an object, just as the Sun's radiation warms objects on Earth. An object such as a thermometer placed in space far away from any other source of heat or radiation would eventually become the same temperature as the "temperature of space.")

What the universe was like before its creation, or if that concept has any meaning at all, remains to be answered. Many people think of the big bang as the beginning of time, so nothing could have possibly come before. This sounds almost too strange to be believed, but perhaps the strangeness arises solely from the limitations of the human mind. People also found it difficult to believe consequences of the special theory of relativity such as time dilation and length contraction, as described in the previous chapter, because these effects had never been encountered before. The creation of the universe was undoubtedly attended by a number of situations and effects never to be encountered by people in their everyday lives.

How the universe evolved from a singularity to its present state is an active subject of research. Perhaps the universe is like a black hole—and some people have suggested it actually is one. Although physicists can piece together some of the universe's evolution using astronomical observations such as the cosmic background radiation, there is a period of time at the very beginning that remains a mystery. The universe was extremely hot and dense, and the laws of physics, at least as they are presently understood, fail to be valid at these extremes. Physics can describe events beginning about 10^{-35} seconds after the big bang—an incredibly short interval of time—but not before.

Over time the universe cooled and expanded. The details of this process are not always clear, but eventually particles such as quarks and electrons condensed, and then as the temperature dropped even more, quarks combined to make protons and other particles. No one knows what happened to any antimatter, whether any appeared at all or whether all of it became annihilated after touching matter, which for some reason may have been created with slightly more abundance than antimatter. Some 300,000 years after the big bang, the temperature had cooled enough for protons to capture electrons, forming hydrogen atoms.

Gravitational attraction caused clumps of matter to appear, although the force of gravitation was clearly not powerful enough to keep the universe from expanding. The clumps became stars, and the congregation of stars formed galaxies. Galaxy formation presents a problem because the universe was probably uniform in the beginning. Physicists believe the answer to this riddle was that a period of rapid expansion called inflation occurred soon after the big bang, leading to variation in densities that would go on to become stars and galaxies.

Verifying what happened in the past is difficult, but astronomers have an advantage. Looking into space is the same as looking back into time. The speed of light is constant, and the universe is so vast that even light takes a long time to travel the distances between stars and galaxies. Astronomers use the term *light-year* as a unit of distance equal to the distance that light travels in a year—approximately 5.9 trillion miles (9.4 trillion km). A light-year is amazingly long, yet the disk (the thickest part) of the Milky Way galaxy is about 100,000 light-years in diameter. The closest

This is an image of a galaxy called the Sombrero Galaxy, for its resemblance to a Mexican hat. *(STScI/NASA/AURA)*

neighboring galaxy, a small one called Canis Major dwarf galaxy, is 25,000 light-years from Earth. (The Canis Major dwarf galaxy lies outside of the Milky Way's disk, but just barely.) Since light takes 25,000 light-years to reach Earth from this galaxy, people on Earth are viewing it as it existed 25,000 years in the past. The farther in space that astronomers can see, the farther back in time they can go. The range goes from just a few years for some of the closer astronomical objects to billions of years for some of the most remote—a length of time going back to the early stages of the universe.

Supernova, Pulsar, and Quasar

The time that light requires to travel the vast distances between stars means that events astronomers observe in space happened years ago. A star seen by Chinese, Arabs, and Native Americans in 1054 suddenly became so bright that it was even noticeable during the day for several weeks. The star was visible at night for several years, though with diminishing brightness, and gradually disappeared. This observation was one of the first recorded instances of an exploding star called a supernova, which briefly becomes tremendously bright, billions of times brighter than the Sun. But the star seen in 1054 did not begin to explode in that year—it is located about 6,300 light-years away, so the event actually took place 6,300 years earlier. It took that long for the light to reach Earth, only then revealing to people on this planet what had happened so far away and so long ago.

A nova is a star that increases in brightness—*nova* means *new star,* for sometimes the increase in brightness makes a star visible where it was too dim to be seen before. But the brightness increase of a supernova is much more dramatic and in many cases is caused by a catastrophic death of the star. This can happen for stars that are at least five times more massive than the Sun. As discussed in chapters 1 and 4, stars maintain a balance throughout most of their lifetimes between the gravitational attraction of their great mass and the outward pressure of their intense heat and radiation. As a star uses up all the nuclei participating in its initial stage of fusion reactions—most of which convert hydrogen into helium—the outward pressure decreases. Although the outer layers of the star swell, gravity causes the central core of the star to shrink. In the process, the center of the star gets hotter, driving the temperature up enough to start a new set of fusion reactions. These reactions fuse heavier nuclei, making elements such as magnesium, sulfur, calcium, and iron. But eventually this activity also ceases as the remainder of small nuclei is used up, and the star has no way to balance the compressing force of gravitation.

As the final collapse begins, the shrinking of the core squeezes the atoms tightly together. Then the compressed core expands as the atoms push away from each other, similar to the bounce of a dropped ball. This bounce pushes away the outer portion of the star with tremendous force, heating the material and sending it streaming out into space. The heat is so intense that several new kinds of nuclear reactions occur, forming heavy elements such as lead, gold, and mercury. Supernova events are the primary source of all the elements in the universe that are heavier than iron—much of the material of Earth and in human bodies was "cooked" in supernova explosions.

There is another type of supernova, occurring not at the end of a star's lifetime but when a star accumulates too much matter from another nearby star or dust cloud. But the explosion is similar, with the same kind of effects.

After the bounce that blows off the outer portion of the star, the core settles down. The density becomes high enough that electrons and protons are squeezed together, forming neutrons. This process is the opposite of radioactive decay, described in chapter 1. A huge number of neutrinos created by the process escape, but the neutrons remain. In some supernova events, the core's collapse stops at this point—gravity acting on the small core is incredibly strong, but the tightly packed neutrons resist any further contraction. The result is called a neutron star. These stars do not shine with light generated by fusion, and although they have more mass than the Sun, their diameter is often less than 12 miles (20 km). Having a density this high—so much material in such a little volume—gives a neutron star gravity that is billions of times greater than Earth is.

A marble on a neutron star would weigh as much as a mountain on Earth.

British astronomer Anthony Hewish and his student Jocelyn Bell Burnell discovered an interesting feature of spinning neutron stars in 1967. Neutron stars not only have huge gravity, but also they have tremendous magnetic fields, around a trillion times stronger than Earth's magnetic field. Such strong magnetic fields sweep up charged particles, producing columns around the magnetic poles. These columns produce an intense beam of radiation. As on Earth, the magnetic poles are not aligned to the axis of rotation. (Earth's magnetic poles are not at the same location as its North Pole and South Pole, and the same is usually true of a rotating neutron star). Rotation of the neutron star causes the beam to sweep around, similar to the beam of a lighthouse. Astronomers observe this beam as a periodic pulse of radiation, and they gave the name *pulsar* to these neutron stars. Pulsars are **pul**sating st**ars.**

The 1054 supernova seen on Earth produced a pulsar located in the Crab Nebula, one of the better-known supernova remnants. Consisting of gas and dust, along with a pulsar in the middle, the Crab Nebula got its name from a drawing made by an astronomer in the middle of the 19th century—apparently it looked something like the animal, although the resemblance was probably due to the limits of the telescopes at the time. The pulsar rotates in excess of 30 revolutions a second, an amazing rate for a massive object, even though it is quite small.

For stars such as the Sun that do not have much mass, the end involves a series of expansions and contractions but nothing as dramatic as a supernova. The Sun will not go out with a tremendously bright explosion, and it will not produce a neutron star. At the end of its lifetime, the Sun will become what is known as a white dwarf, contracting to about the size of Earth. Although it will have a diameter of about 100 times less than its size today, the density will not approach that of neutron stars. (And the Sun has five or so billion years left to live.)

Stars that have about 10 to 12 or more times the mass of the Sun have too much material for neutron stars to form. Some of

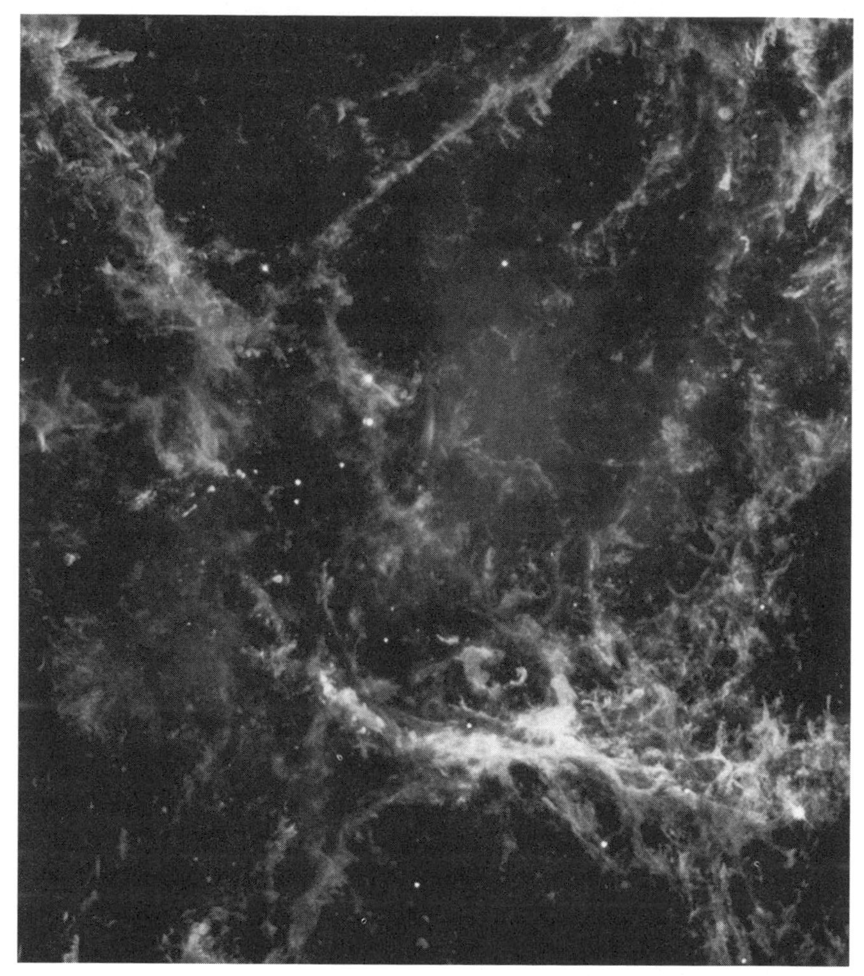

The Crab Nebula is the remnant of the supernova seen on Earth in 1054. *(STScI/NASA/AURA)*

the mass and energy escapes in the supernova explosion, but the remaining matter in the core experiences too much gravitational contraction for neutrons to withstand. As discussed in the previous chapter, and in accordance with the general theory of relativity, a black hole forms (at least in theory). Black holes are so dense that light cannot escape their gravity, but the mass of black holes remains and can attract bits of gas or dust in space, accelerating and heating these particles to the point at which they emit high-energy radiation. As mentioned earlier, astronomers have found many of these telltale signs of black holes throughout the universe, including a massive black hole believed to hide in the center of the Milky Way galaxy. The center of the galaxy has a large concentration of stars, and the black hole may have grown to its present size by gobbling up some of the closer ones.

Black holes may also be an essential component of a mysterious group of objects known as quasars. Quasars are starlike (**quasi**-stell**ar**) objects, appearing as faint stars in telescopes. But the light from these objects has a large redshift, indicating vast distance. To be visible from so far away, these bodies must emit a huge amount of radiation. Some quasars vary rapidly in brightness, a feature typically found in a small object; changes in the brightness of a large object would take time, since light requires time to travel from one end to the other. Thought to be roughly the size of the solar system, quasars manage to shine with as much light as an entire galaxy!

More than 50,000 quasars are known, most farther than 1 billion light-years and some as far as 12 billion light-years away. Many people believe these bright objects are made of the black holes of "active galaxies"—galaxies that emit more light than just the output of their stars. These galaxies may contain an extremely large black hole into which matter is falling, producing a gigantic quantity of radiation. Varying brightness may occur as more or less matter is sucked into the black hole.

Quasars at a distance of 12 billion light-years provide a view of the universe when it was young—light took 12 billion years to arrive, so astronomers on Earth are looking at events occurring soon after creation. Some people think that many or perhaps all galaxies underwent a phase of similar activity in the early universe, a kind of energetic youthfulness before settling down into comfortable middle age.

Theory of Everything

Despite its frenzied activity, the early universe may have been simpler than it is today. At high enough energies, the difference between electromagnetism and the weak force disappears—they have the same properties. This result comes from particle physics, the high-energy physics described in chapter 3 that generates collisions with enough energy density to mimic the early universe. Accelerators cannot produce the same amount of energy as that in the early universe, which is immensely beyond their capacity, but the experiments manage to create events on a small scale involving tremendous amounts of energy. The experiments show that although electromagnetism and the weak force are distinct in ordinary situations, they are the same in high-energy events. Physicists call the combination the electroweak interaction (or force), a unification of **electro**magnetism and the **weak** force.

Unifying, or tying together different laws and theories of physics, is appealing. Condensing a large number of observations or laws into a single statement is elegant, simple, and satisfying. A reduction in the complexity of physics has long been a goal of physicists: in the 17th century, Sir Isaac Newton explained the fall of apples and the orbit of Earth with the same idea—the universal law of gravitation—and in the 1860s James Clerk Maxwell unified electricity and magnetism into a single theory, electromagnetism.

Physicists working today would like to unify electromagnetism, the weak force, and the strong force into a single theory, known as the grand unified theory. Most physicists believe that these forces were unified in the early universe, splitting apart only after the universe expanded and cooled. Several different theories have been proposed, but they are complicated and are difficult to test because the unification only occurs at exceptionally high energies, currently at or beyond the upper limits of particle accelerators.

Grand unified theories are also dissatisfying because they are not as grandly unified as their name suggests. These theories do not include gravity, which a truly unifying theory must. At the very beginning of the universe there may have been only a single force, combining electromagnetism, gravitation, and the strong and weak forces. But proving this assertion may require recreating the incredible temperatures and densities that prevailed shortly after the big bang.

An ultimate theory would unify everything—quantum mechanics (the topic of chapter 2), the Standard Model (chapter 3), relativity theory (chapter 4), along with all forces and interactions, including the nuclear reactions discussed in chapter 1. Some unification is already evident: Electromagnetism and the weak force have a satisfactory unification, the electroweak force mentioned earlier; the Standard Model incorporates quantum mechanics; and nuclear reactions are understandable from Einstein's mass-energy equation along with ideas from quantum mechanics. But ideally, a theory of everything would unite all the laws of physics into a single concept or equation.

Despite recent progress, a theory of everything faces a considerable obstacle. Quantum mechanics and the general theory of relativity are not compatible, and no amount of modification would appear likely to make them fit together. The biggest problem is that there is no place in general relativity for some of the foundations of quantum mechanics, such as Heisenberg's uncertainty principle. Quantum mechanics deals with probabilities, as described in chapter 2, but Einstein's general theory of relativity does not.

Because of this incompatibility, one of these two theories must be wrong. Although both theories have a tremendous amount of experimental support, one (or perhaps both) must be limited. Such limitations exist in Newton's laws, which apply only to a limited range of situations such as slow ("nonrelativistic") speeds and large objects where quantum mechanics is not important. Many people believe that general relativity is the more likely of the two theories to be limited, since gravitational forces normally encountered in the world and in physics experiments are fairly weak. When gravity is very strong, as it would have been in the early universe with its small volume and high density, the effects of quantum mechanics would be apparent.

One theory generating a lot of excitement involves curious objects called strings. Rather than dealing with waves or particles, this theory proposes to explain all of physics with the behavior and interactions of fundamental units consisting of vibrating filaments or strings. String theory might be capable of resolving the mystery of wave-particle duality, mentioned in chapter 2, as well as the incompatibility of quantum mechanics and general relativity. There are several versions of string theory, each requiring complicated mathematics and objects or concepts that are difficult or impossible at present to test. Experimental support is lacking for string theory, so the theory remains an interesting but speculative (unproven) idea.

The Fate of the Universe

Describing all of physics with a single theory or equation would have a satisfying simplicity, but it might never be possible. The universe and its forces, interactions, and objects may be too complex to be summarized in a single, grand statement. Events such as the creation of the universe and its evolution were undoubtedly governed by forces and objects far removed from those that people normally encounter, perhaps even beyond the scope of human perception and imagination.

But the fate of the universe would seem to be an answerable question. The universe will keep expanding or it will not, and it might even one day begin to contract. Gravitational attraction is critical, and the forces exerted by gravitation depend on how much matter the universe contains. A sufficient amount of mass would generate enough gravitation to halt the universe's expansion after a certain period of time and perhaps draw it back inward.

Measuring the size of the universe and the amount of matter it contains is a difficult endeavor. In the process of gathering this information, a surprising result emerged—the expansion of the universe appears to be accelerating. The essential observations came from powerful telescopes such as the *Hubble Space Telescope*, launched in 1990, and other satellites designed to detect distant objects and to measure the cosmic background radiation. Data emerging in recent years about the big bang and the present state of the universe suggest that the expansion is neither constant nor slowing down; rather, it is speeding up.

An accelerating expansion is hard to explain. When something explodes, the fragments fly away at high speeds, but they do not gain any more energy after the explosion—they do not have engines to push them, and the only accelerating force was the initial blast. In Earth's atmosphere, fragments slow down due to air resistance, but in the airlessness of space they would maintain the velocity given to them by the force of the explosion. Even in space, though, the pieces would not speed up. Yet this is what the expanding universe seems to be doing.

Some people have proposed the existence of a kind of substance or energy that has the opposite effect of gravity, exerting a repulsive, antigravity pressure. This energy is called dark energy ("dark" because it is mysterious and concealed). An accelerating expansion would appear to dispense with the need for considering matter density and gravitational attraction, for dark energy will probably continue to push the universe apart. The result would be an eternal expansion, and the universe would grow larger and less dense until, far in the future, little activity could occur.

General relativity can include a concept like dark energy by an addition of a term in the equations, somewhat like Einstein's cosmological constant except it acts in the other direction, to accelerate the expansion instead of stopping or precluding it. Yet no one knows what this dark energy consists of or how to measure it.

Dark energy and the accelerating expansion, if they should both prove valid and not due to misunderstanding or error, are clear examples that the universe continues to harbor secrets. No theory of everything exists, and perhaps at the present level of knowledge, the time is not quite right to make an attempt at formulating one. There is still a lot to learn.

The universe, though probably not boundless as people once thought it was, has a fantastically large extent, stretching out for billions of light-years. The difficulty with understanding the universe arises at least in part because the imagination of most people, except for that of Albert Einstein and a few other brilliant thinkers, is not nearly as large. Astronomical observations and discoveries such as redshifts, pulsars, and quasars reveal intriguing bits and pieces of information, but the creation and evolution of the universe continues to hold unanswered questions. Cosmology is a branch of physics that has made a promising beginning, but there is much left unfinished.

Conclusion

ENCOUNTERS WITH NEW environments and situations usually lead to exciting and often strange new discoveries. European explorers in the 15th and 16th centuries found America, land that no one had known was there. Physicists in the 20th century began to explore the realm of tiny particles and events involving exceptionally fast speeds and found that the theories and principles they had held true for many years no longer applied. The classical physics of Newton, although still useful in certain domains, gave way to quantum mechanics and relativity theory.

Science usually begins with observations and measurements. Attempts at explaining these observations and measurements form the basis for theories, which must be tested by further observations and measurements. Sometimes a theory comes first, as was the case with many of Albert Einstein's ideas. Either way, before anyone can put much faith in a theory, experimenters must test the logical consequences and predictions of the theory. This process is the essence of science. But physics and physicists have struggled in recent times because of it. Beginning with the 20th century and continuing through today, the frontiers of physics often involve incredibly large or small objects and fantastically high speeds. These experiments are almost always expensive to conduct.

Sometimes, physics experiments still involve mostly inexpensive equipment, yet they manage to reveal new and worthwhile discov-

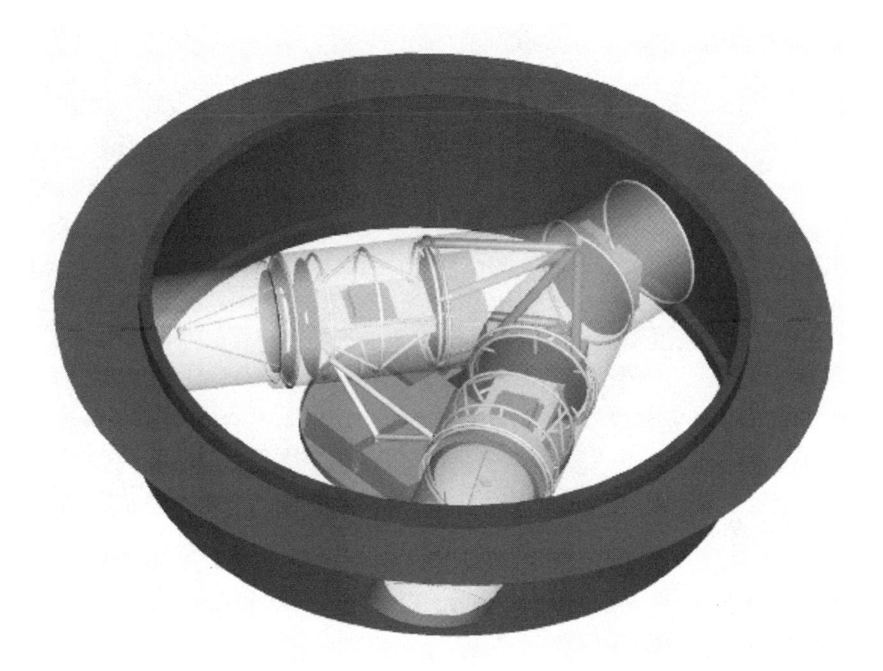

LISA spacecraft will have a Y-shaped structure; working together, the three vehicles form an extended interferometer. *(NASA/JPL-Caltech)*

eries, such as investigations of superconductors. (A superconductor is an electrical conductor that has no electrical resistance, and it is often employed in instruments requiring intense electrical or magnetic effects.) But a lot of modern experiments require equipment such as the gigantic particle accelerators described in chapter 3, costing billions of dollars. Another expensive example is a NASA project, called Laser Interferometer Space Antenna (LISA), to detect gravitational waves. LISA is presently only in the planning stage.

In theory, gravitational waves, as mentioned in chapter 4, originate from masses experiencing gravitational interactions, similar to electromagnetic waves and electromagnetism. The general theory of relativity predicts the existence of gravitational waves, and LISA would be an excellent test for this theory. Finding these waves would not only help support an important theory, but also it would have immediate applications in astronomy, because the waves would offer a useful tool to study the stars and galaxies.

Detecting gravitational waves involves measuring extremely tiny vibrations as they pass through space or matter. Such experiments performed on the surface of Earth are difficult because the planet's movements and vibrations unrelated to gravitational waves mask their presence. The plan for LISA includes three spacecraft to be launched from Earth in rockets. The spacecraft will be identical, with each containing a laser, a telescope, and test masses to provide a gravitational reference sensor. LISA needs three spacecraft because they will position themselves in space to form a huge equilateral triangle—a triangle with equal sides and angles—with each side having a length of 3,125,000 miles (5,000,000 km). In this configuration, the spacecraft will create a large-scale interferometer, similar to that used by Michelson and Morley (described in chapter 4). LISA's equipment is remarkably sensitive, and the goal of the project is to detect a change in the distance between the spacecraft as gravitational waves ripple through the triangle.

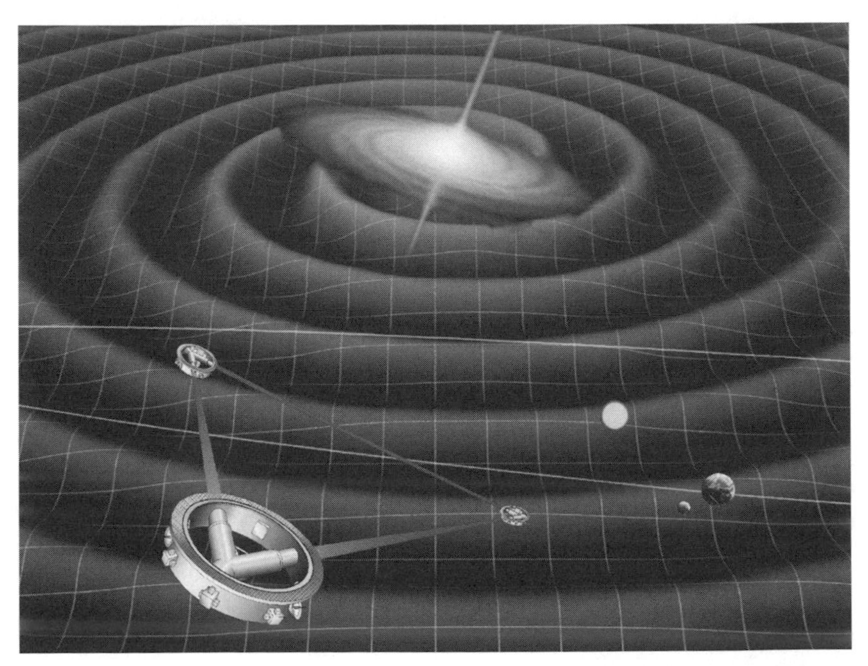

LISA's mission is to detect gravitational waves rippling through space. *(NASA/JPL-Caltech)*

LISA exists only on paper. Plans call for a launch in 2015 or later, but budgets are always a concern. Money for LISA may be diverted to other programs, pushing back the launch date or perhaps canceling the project entirely. Scientific projects funded by government agencies and universities are no strangers to cancellation. A particle accelerator called the Superconducting Super Collider (SSC) in Texas would have permitted some of the highest energies ever achieved in particle physics, but escalating costs led to its demise in 1993, even after construction had already begun and more than a billion dollars had been spent.

Experiments fuel advances in any science such as physics, whether the aim is to make a unique set of observations or to test specific predictions of a theory. Ideas without experimental support are of limited use; string theory, the startling and potentially great advance described briefly in chapter 5, is hard to test and, therefore, hard to believe. The future for the branches of physics discussed in *Particles and the Universe* offers hope of explaining some of the most profound scientific mysteries, yet the hurdles that must be overcome are not just intellectual but also economic.

Albert Einstein used only pen, paper, and his imagination to change the face of physics. But even the ideas of a brilliant physicist such as Einstein cannot be accepted without experimental tests. Perhaps the brightest future for physics will be achieved as gifted thinkers spend time not only on fresh new ideas, but also on the means by which theories can be tested within limited budgets.

SI Units and Conversions

Unit	Quantity	Symbol	Conversion
Base Units			
meter	length	m	1 m = 3.28 feet
kilogram	mass	kg	
second	time	s	
ampere	electric current	A	
Kelvin	thermodynamic temperature	K	1 K = 1°C = 1.8°F
candela	luminous intensity	cd	
mole	amount of substance	mol	
Supplementary Units			
radian	plane angle	rad	π rad = 180 degrees
Derived Units (combinations of base or supplementary units)			
Coulomb	electric charge	C	
cubic meter	volume	m^3	1 m^3 = 1,000 liters = 264 gallons
farad	capacitance	F	
Henry	inductance	H	

Unit	Quantity	Symbol	Conversion
Derived Units (continued)			
Hertz	frequency	Hz	1 Hz = 1 cycle per second
meter/second	speed	m/s	1 m/s = 2.24 miles/hour
Newton	force	N	4.4482 N = 1 pound
Ohm	electric resistance	Ω	
Pascal	pressure	Pa	101,325 Pa = 1 atmosphere
radian/second	angular speed	rad/s	π rad/s = 180 degrees/second
Tesla	magnetic flux density	T	
volt	electromotive force	V	
Watt	power	W	746 W = 1 horsepower

UNIT PREFIXES

Prefixes alter the value of the unit.

Example: kilometer = 10^3 meters (1,000 meters)

Prefix	Multiplier	Symbol
femto	10^{-15}	f
pico	10^{-12}	p
nano	10^{-9}	n
micro	10^{-6}	μ
milli	10^{-3}	m
centi	10^{-2}	c
deci	10^{-1}	d
deca	10	da
hecto	10^{2}	h
kilo	10^{3}	k
mega	10^{6}	M
giga	10^{9}	G
tera	10^{12}	T

Glossary

alpha particles two protons and two neutrons (a helium nucleus) bound together and emitted in certain types of radioactive processes

annihilation the transformation of matter and antimatter, when they meet, into energy

antimatter matter composed of antiparticles, resembling matter but with some of its properties, such as electric charge, opposite in sign

antiparticles antimatter "twins" that exist for particles; for example, the electron's antiparticle is the positron, and the proton's antiparticle is the antiproton

atom the smallest unit of a chemical element, composed of a nucleus surrounded by electrons

atomic number the number of protons in an atom's nucleus

beta particles an electron emitted during radioactive decay

big bang the explosion that created the universe

black hole a small object with so much matter that it creates a gravitational field that is strong enough to prevent anything, including light, from escaping

chain reaction a series of reactions in which one reaction initiates the next

classical physics laws and equations of physics discovered by Sir Isaac Newton and other physicists before the development of quantum mechanics

critical mass in nuclear reactions, a quantity of material sufficient to support a chain reaction

deoxyribonucleic acid (DNA) important molecules in the body that carry inherited information and can be damaged by radiation

determinism the belief that events have specific and identifiable causes instead of occurring by chance

DNA *See* DEOXYRIBONUCLEIC ACID

electromagnetic radiation *See* ELECTROMAGNETIC WAVE

electromagnetic wave a propagating electromagnetic disturbance whose frequency determines its type, such as light, radio waves, and gamma rays; can also behave as particles called photons

electromagnetism describes forces and interactions among charged particles

electron a negatively charged particle. The outer portion of atoms is composed of electrons

electron volt a unit of energy equal to the energy gained by an electron accelerated by 1 volt of electricity

energy the capacity or the potential for motion; in terms of physics, energy is the ability to do work—to apply a force and move an object over some distance

eV *See* ELECTRON VOLT

fallout radioactive material from a nuclear explosion that may be carried long distances by winds before descending to the ground

field a region of space in which a force acts

fission the splitting of an atomic nucleus into two or more parts

fusion the fusing (joining) of nuclei (plural of *nucleus*) to make a heavier nucleus

gamma ray a high-frequency, high-energy electromagnetic wave (or photon)

general theory of relativity discovered by Einstein and based on the notion that physics should be the same for all observers, this theory describes gravitation in terms of geometry rather than as a force

gravitation, force of the attraction between matter, the weakest of the four forces (the other forces are electromagnetism, weak force, and strong force)

half-life the time interval in which half of a radioactive substance decays

Heisenberg's uncertainty principle a scientific principle stating that certain pairs of measurements, such as momentum and position or energy and time, cannot be made with perfect accuracy at the same time; the more accurate one of the pair is known, the less accurate the other must be

Hertz a unit of frequency—the rate at which a periodic event occurs—equal to one cycle per second

ion a charged particle

ionization the process of removing or separating charges to create charged particles

isotopes any of two or more species of atoms of a chemical element having the same number of protons and chemical properties as the element, but a different number of neutrons

light-year the distance light travels in one year, equal to approximately 5.9 trillion miles (9.4 trillion km)

mass the amount of matter a body contains, which affects the body's reaction to a force such as gravitation—more mass means less acceleration for a given force

momentum a measure of motion, defined in classical physics as the product of mass and velocity

NASA *See* NATIONAL AERONAUTICS AND SPACE ADMINISTRATION

National Aeronautics and Space Administration (NASA) the United States government agency devoted to space technology and exploration

neutron electrically neutral particle having a slightly greater mass than the proton

nuclear reaction processes that transform atomic nuclei and release energy in the form of particles and electromagnetic radiation

nuclear reactors devices that harness the energy of nuclear reactions to generate electricity

nucleon a particle in the nucleus—a proton or a neutron

nucleus the central portion of an atom containing the protons and neutrons

photon particle of light or, in general, electromagnetic radiation

plasma a gas consisting of ions

positron an antimatter particle, the antiparticle to the electron

probability the odds or chances that an uncertain event will occur

protons the positively charged particles and components of an atom's nucleus

quantum mechanics the theory governing the behavior of small particles

quarks the fundamental particles composing protons, neutrons, and other particles, bound together tightly by the strong force

radiation the energy propagating across space, carried by waves or particles

radioactive subject to nuclear transformations known as radioactive decay

radioactive dating determining the age of a substance by measuring how much a radioactive component has decayed

radioactive decay the transformation of a nucleus into another, usually accompanied by the emission of particles or radiation

redshift the decrease in frequency of light toward the lower (red) end of the spectrum

special theory of relativity discovered by Albert Einstein and based on the notion that physics should be the same for all observers moving at constant velocity, this theory describes how space and time measurements relate to motion

spectrum a range of frequencies

Standard Model a widely accepted theory classifying the known particles and describing the forces and interactions among them

stochastic based on probability, as opposed to determinism

strong force a powerful but short-range force responsible for holding protons and neutrons together in the nucleus; at its most basic level, it operates on quarks, the components of protons, neutrons, and similar particles

strong nuclear force *See* STRONG FORCE

supernova the violent explosion of a star, accompanied by the production of heavy elements and a brief and dramatic increase in light and other radiation

wavelength the distance between two crests (peaks) of a wave
weak force a short-range force, about a billion times weaker than the strong force, involved in radioactive decay
weak nuclear force *See* WEAK FORCE

FURTHER READING AND WEB SITES

BOOKS

Barnes-Svarney, Patricia L., and Michael R. Porcellino. *Through the Telescope: A Guide for the Amateur Astronomer.* New York: McGraw-Hill, 2000. This is good reading material for backyard astronomers who need some tips getting started.

Calle, Carlos I. *Superstrings and Other Things: A Guide to Physics.* Bristol: Institute of Physics, 2001. Calle explains the laws and principles of physics in a clear and accessible manner.

Einstein, Albert. *Relativity: The Special and General Theory.* New York: Three Rivers Press, 1995. First published in 1920, this edition reprints the words of the discoverer of relativity and one of the world's greatest physicists as he explains his ideas in nontechnical language.

Feynman, Richard P. *QED: The Strange Theory of Light and Matter.* Princeton, N.J.: Princeton University Press, 1986. Feynman, who won the 1965 Nobel Prize in physics, had a remarkable ability to express even the most advanced physics in understandable language and ideas. In this book he explores quantum electrodynamics (QED), a theory that combines quantum mechanics with electromagnetism.

Gribbin, John. *In Search of Schrödinger's Cat: Quantum Physics and Reality.* New York: Bantam, 1984. Quantum mechanics is a difficult subject, but Gribbin does a good job of explaining the strange aspects of the theory.

Hawking, Stephen, and Leonard Mlodinow. *A Briefer History of Time.* New York: Bantam Dell, 2005. This excellent book, coauthored by Stephen Hawking, one of the world's leading physicists, introduces the reader to cosmology, Einstein's theories of relativity, and the history of the universe in an accurate but simple manner.

Henderson, Harry. *Nuclear Physics.* New York: Facts On File, 1998. Telling the story of the development of nuclear physics from a broad perspective, this book focuses on the work of Marie and Pierre Curie, Ernest Rutherford, Niels Bohr, Lise Meitner, Richard Feynman, and Murray Gell-Mann.

Mackintosh, Ray, Jim Al-Khalili, Björn Jonson, and Teresa Peña. *Nucleus: A Trip into the Heart of Matter.* Baltimore, Md.: Johns Hopkins University Press, 2001. Beautifully illustrated and well written, this book examines the forces and structure of the atomic nucleus.

Oerter, Robert. *The Theory of Almost Everything: The Standard Model, the Unsung Triumph of Modern Physics.* New York: Pi Press, 2005. This book delves into the Standard Model of particle physics, which, as the name of the book suggests, is a theory that explains a great deal of physics except the force of gravitation.

Parker, Barry. *Einstein's Brainchild: Relativity Made Relatively Easy.* Amherst, N.Y.: Prometheus Books, 2000. An excellent writer, Parker explains Einstein's relativity theory accurately yet comprehensibly.

Singh, Simon. *Big Bang: The Origin of the Universe.* New York: HarperCollins, 2004. The evolution of the universe and the evolution of human knowledge of this event are complex subjects. Here is an accessible volume for readers who want to explore the complexities in more detail.

Trefil, James. *From Atoms to Quarks: An Introduction to the Strange World of Particle Physics.* New York: Doubleday, 1994. An accessible account of the current state of particle physics and how particle physicists do their research.

WEB SITES

American Institute of Physics. "Physics Success Stories." Available online. URL: http://www.aip.org/success/. Accessed on May 14, 2006. Examples of how the study of physics has impacted society and technology.

American Physical Society. "Physics Central." Available online. URL: http://www.physicscentral.com/. Accessed on May 14, 2006. A collection of articles, illustrations, and photographs explaining physics and its applications, and introducing some of the physicists who are advancing the frontiers of physics even further.

CERN. "Antimatter: The Mirror of the Universe." Available online. URL: http://livefromcern.web.cern.ch/livefromcern/antimatter/index.html. Accessed on May 14, 2006. Developed by CERN, a large research center in Europe that is devoted to particle physics, this set of tutorials explores the history, fundamentals, and production of antimatter.

Cornell University. "Curious About Astronomy? Ask an Astronomer." Available online. URL: http://curious.astro.cornell.edu/. Accessed on May 27, 2006. This Web site invites visitors to submit questions about astronomy and also provides a list, with answers, of previously asked questions.

Exploratorium: The Museum of Science, Art and Human Perception. Available online. URL: http://www.exploratorium.edu/. Accessed on May 14, 2006. An excellent web resource containing much information on the scientific explanations of everyday things.

Fermilab Education Office homepage. Available online. URL: http://www-ed.fnal.gov/ed_home.html. Accessed on May 14, 2006. Fermilab maintains a valuable collection of information and tutorials on particles physics in general and the research conducted at their facility.

Friedman, S. Morgan. "Albert Einstein Online." Available online. URL: http://www.westegg.com/einstein/. Accessed on May 14, 2006. A collection of links to the writings, quotes, pictures, biographies, and opinions of Albert Einstein.

HowStuffWorks, Inc., homepage. Available online. URL: http://www.howstuffworks.com/. Accessed on May 14, 2006. Contains a large number of articles, generally written by knowledgeable authors, explaining the science behind everything from computers to satellites.

Lawrence Berkeley National Laboratory (Particle Data Group). "The Particle Adventure." Available online. URL: http://particleadventure.org/particleadventure/. Accessed on May 14, 2006. These highly recommended Web pages offer an engaging and easily understandable account of particles, antiparticles, accelerators, detectors, and much more.

National Aeronautics and Space Administration. "Cosmology 101." Available online. URL: http://map.gsfc.nasa.gov/m_uni.html. Accessed on May 14, 2006. This tutorial includes discussions of the foundations, observational tests, and limitations of the big bang theory, along with the subsequent evolution, shape, and fate of the universe.

———. "Imagine the Universe!" Available online. URL: http://imagine.gsfc.nasa.gov/. Accessed on May 14, 2006. Written for students, this site is full of news and information on astronomy, physics, and the contributions of space exploration and technology to these subjects.

Nave, Carl R. "HyperPhysics Concepts." Available online. URL: http://hyperphysics-phyastr.gsu.edu/hbase/hph.html. Accessed on May 9, 2006. This comprehensive resource for students offers illustrated explanations and examples of the basic concepts of all the branches of physics, including quantum physics.

Nuclear Energy Institute (NEI) homepage. Available online. URL: http://www.nei.org/. Accessed on May 14, 2006. NEI members include companies involved in the maintenance and operation of nuclear power plants and companies involved in the field of nuclear medicine. This institute helps set policies affecting the industry, and its Web page includes news and basic information on all aspects of nuclear energy.

Stanford Linear Accelerator Center (SLAC). "Explore the Virtual Visitor Center." Available online. URL: http://www2.slac.stanford.edu/vvc/Default.htm. Accessed on May 14, 2006. This Web site offers a tour of SLAC and its research, featuring pictures and discussions of equipment and experiments.

PERIODIC TABLE OF THE ELEMENTS

Key: 3 (Atomic number) — Li (Symbol) — 6.941 (Atomic weight)

1 IA	2 IIA	3 IIIB	4 IVB	5 VB	6 VIB	7 VIIB	8 VIIIB	9 VIIIB	10 VIIIB	11 IB	12 IIB	13 IIIA	14 IVA	15 VA	16 VIA	17 VIIA	18 VIIIA
1 H 1.00794																	2 He 4.0026
3 Li 6.941	4 Be 9.0122											5 B 10.81	6 C 12.011	7 N 14.0067	8 O 15.9994	9 F 18.9984	10 Ne 20.1798
11 Na 22.9898	12 Mg 24.3051											13 Al 26.9815	14 Si 28.0855	15 P 30.9738	16 S 32.067	17 Cl 35.4528	18 Ar 39.948
19 K 39.0938	20 Ca 40.078	21 Sc 44.9559	22 Ti 47.867	23 V 50.9415	24 Cr 51.9962	25 Mn 54.938	26 Fe 55.845	27 Co 58.9332	28 Ni 58.6934	29 Cu 63.546	30 Zn 65.409	31 Ga 69.723	32 Ge 72.61	33 As 74.9216	34 Se 78.96	35 Br 79.904	36 Kr 83.798
37 Rb 85.4678	38 Sr 87.62	39 Y 88.906	40 Zr 91.224	41 Nb 92.9064	42 Mo 95.94	43 Tc (98)	44 Ru 101.07	45 Rh 102.9055	46 Pd 106.42	47 Ag 107.8682	48 Cd 112.412	49 In 114.818	50 Sn 118.711	51 Sb 121.760	52 Te 127.60	53 I 126.9045	54 Xe 131.29
55 Cs 132.9054	56 Ba 137.328	57-70 ☆; 71 Lu 174.967	72 Hf 178.49	73 Ta 180.948	74 W 183.84	75 Re 186.207	76 Os 190.23	77 Ir 192.217	78 Pt 195.08	79 Au 196.9655	80 Hg 200.59	81 Tl 204.3833	82 Pb 207.2	83 Bi 208.9804	84 Po (209)	85 At (210)	86 Rn (222)
87 Fr (223)	88 Ra (226)	89-102 ★; 103 Lr (260)	104 Rf (261)	105 Db (262)	106 Sg (266)	107 Bh (262)	108 Hs (263)	109 Mt (268)	110 Ds (271)	111 Rg (272)	112 Uub (277)	113 Uut (284)	114 Uuq (285)	115 Uup (288)	116 Uuh (292)	117 Uus ?	118 Uuo ?

Numbers in parentheses are atomic mass numbers of most stable isotopes

☆ Lanthanoids	57 La 138.9055	58 Ce 140.115	59 Pr 140.908	60 Nd 144.24	61 Pm (145)	62 Sm 150.36	63 Eu 151.966	64 Gd 157.25	65 Tb 158.9253	66 Dy 162.500	67 Ho 164.9303	68 Er 167.26	69 Tm 168.9342	70 Yb 173.04
★ Actinoids	89 Ac (227)	90 Th 232.0381	91 Pa 231.036	92 U 238.0289	93 Np (237)	94 Pu (244)	95 Am 243	96 Cm (247)	97 Bk (247)	98 Cf (251)	99 Es (252)	100 Fm (257)	101 Md (258)	102 No (259)

THE CHEMICAL ELEMENTS

(g) none (c) nonmetallics

element	symbol	a.n.
carbon	C	6
hydrogen	H	1

(g) chalcogen (c) nonmetallics

element	symbol	a.n.
oxygen	O	8
polonium	Po	84
selenium	Se	34
sulfur	S	16
tellurium	Te	52
ununhexium	Uuh	116

(g) alkali metal (c) metallics

element	symbol	a.n.
cesium	Cs	55
francium	Fr	87
lithium	Li	3
potassium	K	19
rubidium	Rb	37
sodium	Na	11

(g) alkaline earth metal (c) metallics

element	symbol	a.n.
barium	Ba	56
beryllium	Be	4
calcium	Ca	20
magnesium	Mg	12
radium	Ra	88
strontium	Sr	38

(g) none (c) metallics

element	symbol	a.n.
aluminum	Al	13
bohrium	Bh	107
cadmium	Cd	48
chromium	Cr	24
cobalt	Co	27
copper	Cu***	29
darmstaduim	Ds	110
dubnium	Db	105
gallium	Ga	31
gold	Au***	79
hafnium	Hf	72
hassium	Hs	108
indium	In	49
iridium	Ir ****	77
iron	Fe	26
lawrencium	Lr	103
lead	Pb	82
lutetium	Lu	71
manganese	Mn	25
meitnerium	Mt	109
mercury	Hg	80
molybdenum	Mo	42
nickel	Ni	28
niobium	Nb	41
osmium	Os****	76
palladium	Pd****	46
platinum	Pt ****	78
rhenium	Re	75
rodium	Rh****	45
roentgenium	Rg	111
ruthenium	Ru****	44
rutherfordium	Rf	104
scandium	Sc	21
seaborgium	Sg	106
silver	Ag***	47
tantalum	Ta	73
technetium	Tc	43
thallium	Tl	81
titanium	Ti	22
tin	Sn	50
tungsten	W	74
ununbium	Uub	112
ununtrium	Uut	113
ununquadium	Uuq	114
vanadium	V	23
yttrium	Y	39
zinc	Zn	30
zirconium	Zr	40

(g) pnictogen (c) metallics

element	symbol	a.n.
arsenic	As*	33
antimony	Sb*	51
bismuth	Bi	83
nitrogen	N	7
phosophorus	P**	15
ununpentium	Uup	115

(g) none (c) semimetallics

element	symbol	a.n.
boron	B	5
germanium	Ge	32
silicon	Si	14

(g) actinoid (c) metallics

element	symbol	a.n.
actinium	Ac	89
americium	Am	95
berkelium	Bk	97
californium	Cf	98
curium	Cm	96
einsteinium	Es	99
fermium	Fm	100
mendelevium	Md	101
neptunium	Np	93
nobelium	No	102
plutonium	Pu	94
protactinium	Pa	91
thorium	Th	90
uranium	U	92

(g) halogens (c) nonmetallics

element	symbol	a.n.
astatine	At*	85
bromine	Br	35
chlorine	Cl	17
fluorine	F	9
iodine	I	53
ununseptium	Uus*	117

(g) lanthanoid (c) metallics

element	symbol	a.n.
cerium	Ce	58
dysprosium	Dy	66
erbium	Er	68
europium	Eu	63
gadolinium	Gd	64
holmium	Ho	67
lanthanum	La	57
neodymium	Nd	60
praseodymium	Pr	59
promethium	Pm	61
samarium	Sm	62
terbium	Tb	65
thulium	Tm	69
ytterbium	Yb	70

(g) noble gases (c) nonmetallics

element	symbol	a.n.
argon	Ar	18
helium	He	2
krypton	Kr	36
neon	Ne	10
radon	Rn	86
xenon	Xe	54
unococtium	Uuo	118

a.n. = atomic number
(g) = group
(c) = classification

* = semimetallics (c)
** = nonmetallics (c)
*** = coinage metal (g)
**** = precious metal (g)

INDEX

ASHLAND COMMUNITY & TECHNICAL COLLEGE
3 3631 1131361 4